Carola Otto

Fusion of Data from Heterogeneous Sensors with Distributed Fields of View and Situation Evaluation for Advanced Driver Assistance Systems

**Forschungsberichte aus der Industriellen Informationstechnik
Band 8**

Institut für Industrielle Informationstechnik
Karlsruher Institut für Technologie
Hrsg.   Prof. Dr.-Ing. Fernando Puente León
         Prof. Dr.-Ing. habil. Klaus Dostert

Eine Übersicht über alle bisher in dieser Schriftenreihe erschienenen Bände
finden Sie am Ende des Buchs.

Fusion of Data from Heterogeneous Sensors with Distributed Fields of View and Situation Evaluation for Advanced Driver Assistance Systems

von
Carola Otto

Dissertation, Karlsruher Institut für Technologie (KIT)
Fakultät für Elektrotechnik und Informationstechnik
Tag der mündlichen Prüfung: 6. Juni 2013
Referenten: Prof. Dr.-Ing. F. Puente León, Prof. Dr.-Ing. C. Stiller

Impressum

Karlsruher Institut für Technologie (KIT)
KIT Scientific Publishing
Straße am Forum 2
D-76131 Karlsruhe
www.ksp.kit.edu

KIT – Universität des Landes Baden-Württemberg und
nationales Forschungszentrum in der Helmholtz-Gemeinschaft

KIT Scientific Publishing 2013
Print on Demand

ISSN 2190-6629
ISBN 978-3-7315-0073-5

Fusion of Data from Heterogeneous Sensors with Distributed Fields of View and Situation Evaluation for Advanced Driver Assistance Systems

Zur Erlangung des akademischen Grades eines

DOKTOR-INGENIEURS

von der Fakultät für

Elektrotechnik und Informationstechnik

des Karlsruher Instituts für Technologie (KIT)

genehmigte

DISSERTATION

von

Dipl.-Ing. Carola Otto

geb. in Altdorf bei Nürnberg

<table>
<tr><td>Tag der mündlichen Prüfung:</td><td>6. Juni 2013</td></tr>
<tr><td>Hauptreferent:</td><td>Prof. Dr.-Ing. F. Puente León, KIT</td></tr>
<tr><td>Korreferent:</td><td>Prof. Dr.-Ing. C. Stiller, KIT</td></tr>
</table>

Acknowledgements

This dissertation is the product of a fruitful cooperation between *Daimler Trucks Advanced Engineering* and the *Institut für Industrielle Informationstechnik* (IIIT) which is part of the Karlsruhe Institute of Technology (KIT).

First and foremost, I would like to thank Professor Dr.-Ing. Fernando Puente León, Head of the IIIT, for his great supervision throughout my PhD project.

I am very grateful to Professor Dr.-Ing. Christoph Stiller from the *Institut für Mess- und Regelungstechnik* for his interest in my work and for accepting to read and comment on the manuscript as second reviewer.

Professor Dr. rer.nat. Friedrich Jondral and Professor Dr.-Ing. Sören Hohmann amiably presided the examination board while Professor Dr. rer.nat. Olaf Dössel kindly accepted to head the jury.

A particularly big share of my gratitude goes to my current and former colleagues in the department *System Applications* of Daimler Trucks Advanced Engineering - Dr.-Ing. Andreas Schwarzhaupt, Dr.-Ing. Jan Wirnitzer, Daniel Hillesheim, Dr.-Ing. Istvan Vegh, Dr.-Ing. Urs Wiesel and Markus Kirschbaum. They gave me the opportunity to work on this topic and supported me with all the required materials. Furthermore, I want to thank my colleagues of the Daimler Research Department for various fruitful discussions and for providing basic components for this work, for example, the camera-based pedestrian detection algorithm or object labeling tools, i.e.: Matthias Oberländer, Dr.-Ing. Clemens Rabe and Dr.-Ing. Michael Munz. I have been able to profit from their experience and advice numerous times during my time as a doctoral student. I would also like to thank all of my Daimler colleagues for the great atmosphere and the welcomed distractions from day-to-day problems.

Furthermore, I would like to express my warmest gratitude to the colleagues at the IIIT in Karlsruhe. I could always count on their support, and I particularly enjoyed the amicable discussions during the summer seminars.

Among the colleagues and friends, I am particularly indebted to Antje Westenberger and Dr.-Ing. Friederike Brendel for their enormous effort when proofreading my manuscript with impressive patience and accuracy.

Moreover, I want to thank all students who contributed with their theses to this work; i.e.: Isabel Thomas, Andreas Pflug, Daniel Penning, Matheus Bianchi Dambros, and Wladimir Gerber.

Last but not least, I would like to thank my closest friends and family, and in particular my mother Elke Otto and my boyfriend Christoph Schmidt, for supporting me throughout the course of this thesis.

Karlsruhe, June 2013   Carola Otto

Kurzfassung

Jährlich sterben mehrere Tausend Verkehrsteilnehmer auf Europas Straßen. Fahrerassistenzsysteme trugen in den letzten Jahren wesentlich zur Reduktion der Straßenverkehrsopfer in der EU bei [1]. Jedoch blieb die Anzahl der Unfälle mit ungeschützten Verkehrsteilnehmern weitgehend konstant [2]. Bei Zusammenstößen zwischen Fußgänger und LKW können schon geringe Kollisionsgeschwindigkeiten zu schweren Verletzungen oder Tod führen. Deshalb ist eine zuverlässige Prävention erforderlich.

Dafür ist zunächst die Erkennung, Vermessung und Verfolgung von Fußgängern im Fahrzeugumfeld erforderlich. Ein neuartiger Sensordatenfusionsansatz zur Fußgängerverfolgung von einem LKW aus wurde entwickelt, implementiert und für die Echtzeitanwendung im LKW parametriert. Der Ansatz basiert auf dem Joint Integrated Probabilistic Data Association Filter (JIPDA) und nutzt Daten von drei Radaren und einer Monokularkamera. Während ein Nahbereichsradar, ein Fernbereichsradar und die Kamera das frontale Umfeld des Fahrzeugs überwachen, liefert ein an der rechten Seite des Fahrzeugs angebrachtes Radar Messungen aus dem Totwinkelbereich. Fußgänger werden in sich nicht überlappenden Sensorsichtbereichen und über einen sensorisch blinden Bereich hinweg mit heterogenen Sensoren verfolgt. Somit werden vorhandene Objektinformationen auch für den sensorisch blinden Bereich verfügbar gemacht und an den sich anschließenden zweiten Erfassungsbereich weitergegeben. Ein Vergleich eines erweiterten Kalmanfilters mit globaler Nächster-Nachbar-Datenassoziation hinsichtlich Detektionsperformanz und Genauigkeit belegt die Vorteile des neu entwickelten Ansatzes.

Neben den Informationen über Objekte im Fahrzeugumfeld ist für die Einschätzung des Gefährdungspotenzials einer Situation, die Vorhersage der Bewegung des eigenen Fahrzeugs wichtig. Es wurde ein neuartiger, robuster Ansatz entwickelt, welcher die Trajektorien des eigenen Fahrzeugs rein auf Basis fahrzeugeigener CAN-Daten in die Manöver Spurfolgen, Abbiegen und Spurwechsel klassifiziert und den zukünftigen Pfad prädiziert. Die Genauigkeit des neuen Prädiktionsansatzes wurde anhand der später tatsächlich gefahrenen Trajektorie bewertet, und die Ergebnisse wurden mit jenen eines etablierten Standardansatzes verglichen. Die Prädiktionsgenauigkeit kann durch den neu entwickelten Ansatz im Vergleich zum Standardansatz signifikant erhöht werden.

Zur Bestimmung der Kollisionsgefahr des eigenen Fahrzeugs mit anderen Verkehrsteilnehmern, insbesondere Fußgängern, werden zunächst die stochastisch erreichbaren Zustandsmengen der Verkehrsteilnehmer entlang ihrer Pfade für Zeitpunkte und Zeitintervalle berechnet. Aus den Überlappungsbereichen der Erreichbarkeitsmengen kann schließlich die Kollisionswahrscheinlichkeit mit einem detektierten Verkehrsteilnehmer als Gefährdungsmaß bestimmt werden. Der Vorteil des neuen Ansatzes im Vergleich zu einem herkömmlichen Ansatz besteht darin, dass sensorische Ungenauigkeiten und Unsicherheiten in der Vorhersage explizit berücksichtigt werden können. Dadurch führt der neue Ansatz in Szenarien, in denen die Unsicherheiten stark realisiert sind, zu stabileren Entscheidungen.

Contents

Abbreviations and Symbols — 207

Bibliography — 221

1 Introduction

1.1 Motivation

Several thousand road users die in traffic accidents on European streets every year. Advanced driver assistance systems (ADAS) contributed to the decreasing number of traffic victims in the European Union during the last years [1]. Systems of active safety like the electronic stability control (ESC), forward collision warning (FCW) towards other vehicles and the lane departure warning (LDW) have shown their benefit in such a way that these systems became legal requirements in commercial vehicles heavier than 3.5 tons in the European Union.

However, the goal of accident-free driving has not been reached yet. Especially, the number of lethal accidents with vulnerable road users stayed about constant throughout the last years [2]. Beside the social cost of road traffic injuries, the economic cost of road crashes and injuries is estimated to be about 2 % of the gross national product (GNP) in high income countries [3]. The negligible mass of pedestrians and cyclists compared to cars is a physical disadvantage in collisions. Extremely precarious situations result if a vulnerable road user collides with a truck. As pedestrians and cyclists are strongly endangered and vulnerable even at small collision speeds, reliable prevention is required. Active safety systems that avoid collisions with vulnerable road users and consider the singularities of trucks and their drivers have to be developed.

Truck drivers usually drive for hours before they enter urban environment to deliver their freight. Operator fatigue after a long trip might be one reason why pedestrians at crosswalks are overlooked. Complex crossing scenarios can be another one, for instance in situations where several pedestrians cross the street at once and others enter the blind spot area of the truck. Full concentration of the driver is required in order to avoid collisions during turning maneuvers. Systems that support drivers in such demanding situations by surveillance of the vehicle's blind spot are desirable.

There are several ADAS using sensor systems for environment perception that warn the driver of a potential collision with another car or even conduct an emergency braking autonomously if the driver does not react.

Yet, systems that react to vulnerable road users are rare and the number of implemented use cases is very limited [4]. A shorter time to market of these systems can be reached if the environment perception uses sensors that already exist at the truck, such as radar sensors for FCW and a monocular camera for LDW. Radars detect cars reliably due to their large cross section of strongly reflecting metal, while they initially have not been designed to detect pedestrians that only provide diffuse reflections. The provided radar cross section (RCS) values and absorption characteristics are similar to those of reflections of rough ground or bushes. Thus, a high false alarm rate and low detection probabilities that are not distributed uniformly have to be taken into account for sensor data fusion for pedestrian tracking. Beside estimation of the state, information about the certainty of the sensor framework is strongly desired for an evaluation of the situation and a consequential application decision. If a system intervenes into the responsibility of the driver to avoid collisions by steering or braking, there must not be any doubt about the criticality of the situation and the existence of detected objects. The dynamics of pedestrians are highly volatile. Pedestrians do not necessarily follow the lane, but may suddenly start to cross it. This makes the situation evaluation (prediction) by an ADAS much more complex than in the context of vehicles driving ahead on the same or adjacent lanes. In a scenario where a pedestrian is located on the roadside, the pedestrian might walk into the driving corridor of the ego vehicle[1] as he might not have recognized the oncoming vehicle (truck) or over-estimated the gap. However, he might just stay on the roadside or walk along the street. Warnings for the driver have to be provided early enough to enable a collision avoidance, although the situation might change completely in the meantime. False alarms decrease the driver's acceptance for the system and therefore need to be suppressed. Hence, an ADAS for pedestrian protection not only requires a reliable detection and tracking algorithm that provides information about spatial uncertainties and the object's probability of existence (PoE) but also a situation evaluation that can properly handle the uncertainties of the pedestrians' motion as well as the ego vehicle's motion.

1.2 Focus and Contributions of the Thesis

This thesis is focused on several aspects of environment perception and situation analysis in ADAS of commercial vehicles for protection of vul-

---

[1] The vehicle hosting the developed system is referred to as ego vehicle in this work.

nerable road users. The main contributions refer to environment perception and situation evaluation.

Environment Perception

A novel sensor fusion approach for pedestrian tracking from a truck is developed, implemented and parametrized for real-time application. The approach is based on the Joint Integrated Probabilistic Data Association (JIPDA) filter and uses detections from radar sensors and a monocular camera. Pedestrians are tracked in distinct fields of view (FOV) and across a sensory blind region using heterogeneous sensors. Thereby, tracks can be confirmed earlier than when using single filters for each FOV. Moreover, object information is made available in the sensory blind region. Information that is only available in one FOV can be exploited in another FOV and enables, for instance, a more reliable object classification in the FOV behind the sensory blind region.

Pedestrian detections are filtered from the reflection points of a laser scanner. The radars' and the camera's spatial inaccuracies and their detection performances are measured dependent on measurement properties and current object states using the pedestrian detections from the laser scanner as reference measurements. An own implementation of the extended Kalman filter (EKF) with global nearest neighbor data association (GNN) enables a comparative evaluation of the performance of the developed approach.

Situation Evaluation

Information about objects in the vehicle's environment represents one basic component to evaluate the risk of a situation. The prediction of the ego vehicle's motion behavior builds the second component. Therefore, a new approach that classifies trajectories of the ego vehicle into the maneuvers turn, lane change and lane following is presented. The approach uses data from CAN (Controller Area Network) only and is based on the Longest Common Subsequence (LCS) method and a Bayesian classifier. The trajectory of the ego vehicle is predicted based on the classified maneuver and adjustable model trajectories. The performance of the developed approach is evaluated in comparison to the ground truth and an established approach.

The prediction of the collision risk with a pedestrian requires the understanding of typical pedestrian behavior. Therefore, pedestrian trajectories

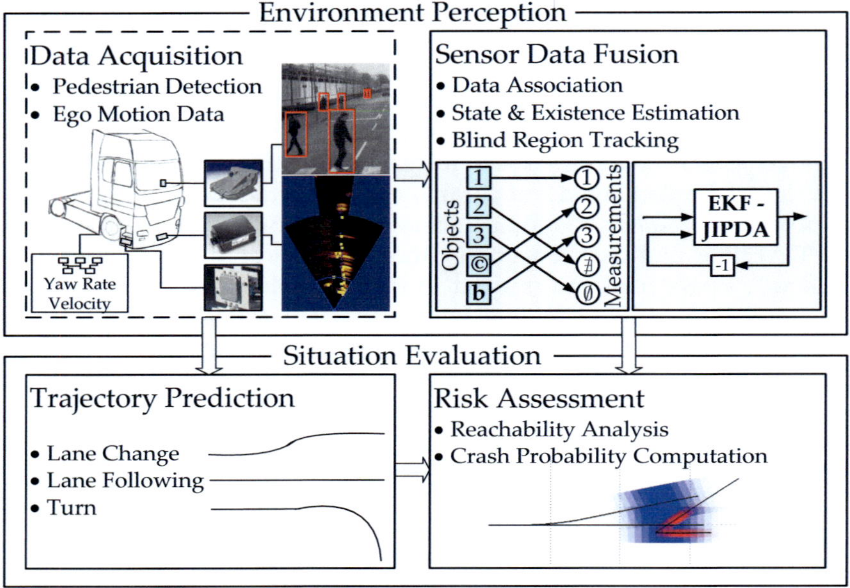

Figure 1.1   Overview of the developed system with its main components.

and pedestrian motion behavior are analyzed in an exploratory study. The results are compared to the results from the literature.

A conditioned crash probability of a pedestrian with the ego vehicle is then computed based on stochastic reachable sets and Markov chain abstraction where the algorithm accounts for uncertainty in the road users' motion and in the existence of a tracked object as well as for temporal and spatial uncertainties.

1.3 Structure of the Thesis

Figure 1.1 depicts the main components of the developed approach. The data acquisition is taken as given. A novel method for pedestrian tracking fusing data from radar sensors and a camera is introduced. For situation evaluation, the trajectories of the road users are predicted and the collision probability of the ego vehicle with a detected road user is computed. The thesis is structured into seven chapters which are shortly described.

Chapter 2 covers the fundamentals of ADAS and their structure. It introduces different categories of ADAS and basic conditions for developing such systems. Subsequently, the fundamentals of environment perception, the measurement principles of applied sensors as well as the background of sensor fusion are described. Moreover, the chapter outlines the basics for situation evaluation in context with vulnerable road users such as typical crash scenarios and typical pedestrian dynamics.

Chapter 3 concentrates on the theoretical foundations that are required for object tracking including different approaches for data association. Furthermore, the basics of classification, discrete event systems and Markov chain abstraction are introduced, as these constitute the required theoretical background for the situation evaluation.

The first part of Chapter 4 refers to related work regarding sensor data fusion using radar and camera, pedestrian tracking, and tracking across sensory blind regions. The following sections present the implementation of two pedestrian tracking approaches using two radar sensors at the front bumper, a monocular camera behind the wind shield, and a radar sensor on the right side of the truck. The two tracking approaches are based on different concepts of data association — GNN and JIPDA. A new procedure is introduced for measuring sensor properties with respect to pedestrian detection. A major contribution is stated by the fact that pedestrians are tracked across a sensory blind region from a moving platform.

Chapter 5 enumerates the most relevant, recent and existing approaches for maneuver classification, trajectory prediction, and threat assessment where the demand for novel approaches is discussed. The chapter is dedicated to a new prediction approach for the ego vehicle's trajectory and the computation of a novel risk measure for collisions between the ego vehicle and pedestrians using stochastic reachable sets and Markov chain abstraction. The results of an own study regarding pedestrian motion behavior in traffic constitute one basis of the stochastic reachable sets.

Chapter 6 compares and discusses the results obtained by the developed approaches to the results of existing and established methods. It is shown that the detection performance of the novel approach for pedestrian tracking based on JIPDA outperforms the GNN-based approach. Especially, turn maneuvers of the ego vehicle can be classified well, so that the prediction of the ego vehicle's path becomes more accurate with the developed approach than with the established approach for this maneuver class. The evaluation of the developed approach for computation of a crash probability of the ego vehicle with another road user indicates that sensory and situational uncertainties should be accounted in every risk assessment approach. The thesis concludes with a summary and an outlook on potential future work in Chapter 7.

2 Advanced Driver Assistance Systems

2.1 Categorization and Regulatory Standards

ADAS have contributed to the decrease of the number of traffic victims during the last years. While passive safety technologies (such as airbags or rear under-run protection devices) become relevant during an accident and mitigate its effects, active safety systems operate continuously in order to avoid potential accidents.

Four different classes of ADAS can be distinguished [5] from the functional perspective. The boundaries are fuzzy and thus, systems may belong to more than one class at once.

Autonomous systems intervene actively in the vehicle dynamics without a situational initiation by the driver. Examples are the anti-lock braking system (ABS) and the ESC, which were introduced into commercial vehicles in the years 1981 and 2001 for the first time [6]. These systems manipulate the actors without the driver's influence. These systems are usually applied when situations are not controllable by any human driver, for instance, braking of single tires, as performed by ESC, might be required to control an otherwise hopeless situation.

Comfort systems accomplish a part of the driving task and should thereby relieve the driver. However, his engagement is required at least in the sense of supervision. The final responsibility for the vehicle control stays with the driver. Thus, the system's behavior needs to be comprehensible and predictable for the driver. Moreover, he shall be able to override any system reaction. This requirement was formulated in the *Vienna Convention on Road Traffic* [7] in 1968, where it reads in article 8, subparagraph 5: *'Every driver shall at all times be able to control his vehicle or guide his animals.'* A prominent example for a comfort system is the adaptive cruise control system (ACC) that controls the speed dependent on the distance and relative speed to preceding objects.

Driver information systems provide information supporting drivers in their control action without direct intervention in the driving behavior. Navigation systems, traffic sign recognition for no-passing zone information or speed limit assistance systems (SLA) or even lane departure warning systems that alert drivers when they start to cross the lane marking without having set the turn signal are some examples of such information systems.

Efficiency enhancing systems are especially relevant in commercial vehicles since one third of the costs for truck-based freight companies is caused by fuel [8]. The systems aim for a vehicle control with optimized energy consumption. This can be performed, e.g., by informing the driver about curves or road gradients ahead for a more predictive way of driving or by directly controlling the gears, the engine speed and the energy distribution in hybrid drives. If the behavior of other road users could be predicted as well, for instance by using car-to-car communication, this knowledge could be exploited not only for collision avoidance functions but also for a more efficient vehicle control.

Computer vision-based ADAS necessitate environment perception modules that are able to deal with the object diversity and environmental conditions of the real world. The required information for an assistance function has to be retrieved reliably in a sufficient FOV. The extent of fulfillment often determines the systems' costs. This usually leads to trade-offs, so that for instance, the number of function use cases is reduced to scenarios where the criticality of the situation can be recognized unambiguously. The availability of dedicated sensors is a key element for the realization of corresponding functions where self-surveillance and self-diagnosis of the components are requirements. The impact of environmental conditions on the system behavior is inevitable, but the system shall detect its capabilities within the current situation and provide a confidence level or plausibility value of its information. For example, a camera with a soiled lens should not provide data based on the soiled spots, but report its state.

The imperfection of different components of an ADAS is inherent in the system. Suited system architectures and self-surveillance are required to guarantee a safe system behavior with uncertain components. The *ISO26262* [9] (Road Vehicles — Functional Safety) is a guideline for the required careful development and production process of these systems. It aims at functional safety but does not set requirements for the performance of a function. Systematic errors, e.g., due to wrong specifications or wrong implementation, have to be eliminated. System stability as well as system robustness in case of component failure, should be guaranteed.

This ISO standard assigns a degree of criticality to a system based on the following three criteria: severity of a possible injury due to a system failure, controllability of the system and the situation by the driver, and failure frequency of the system. The required analysis has to be performed based on a set of imaginable scenarios that the system could undergo. The system obtains one of four **Automotive Safety Integrity Levels** (ASIL) after the system classification based on these criteria. Conventional quality standards are sufficient for the development and production process of systems with the lowest level ASIL A. Systems that have assigned the highest integrity level (ASIL D) need to satisfy demanding criteria regarding redundancy of data and algorithms, documentation, formal verification, measurement and component tolerances, self-diagnosing capabilities, design review, and component tests among others.

A driver assistance system can be considered as a closed-loop control system (see Figure 2.1) with a human (driver) and an artificial controller. Good coordination of both controllers is required for cooperative behavior, but a clear segregation between the responsibilities of the human driver and the machine for the driving task is not trivial. Auditive, visual or haptic feedback are solutions to inform the driver that the system will take over control if he fails to react within a certain time interval. The artificial controller aims to follow the set target and bases its decisions on sensor data and information gained through data fusion. The situation evaluation module processes information about the environment and needs to incorporate the driver's wish. It classifies the situation before a decision for a certain system behavior is made and a request is sent to the actuators.

The capability of sensor-based environment perception is still below human abilities, in particular considering the reliability — meaning the correctness of statement that an object exists and the consistency of the estimation result, or the classification performance. However, artificial environment perception (computer vision) provides advantages as well — such as a shorter latency time in the processing chain consisting of perception, decision, and action or a higher metric accuracy. Moreover, computer vision does not suffer from distraction or fatigue.

Section 2.2 describes the basics regarding automotive environment perception. It includes a description of the sensors that have been utilized in this work as well as an overview over different methods and motivations to fuse their data.

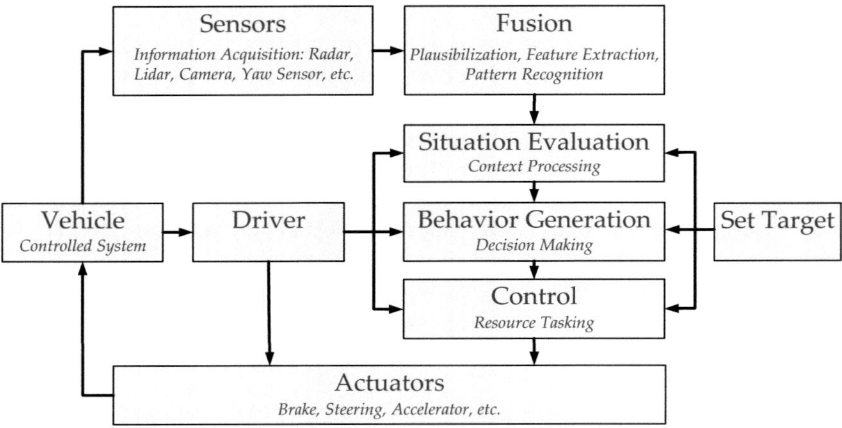

Figure 2.1 Schematic of control loop in ADAS: Control of driver overrides ADAS control.

2.2 Automotive Computer Vision

The environment perception module that is presented in this work is capable to fuse data from different sensor modules in a generic way. Knowledge about usable sensor types and their measurement principles is required to realize an abstract interface, so that this section provides information about sensors that are applicable to automotive computer vision and pedestrian detection.

2.2.1 Sensor Modules

A *sensor module* is defined as a component consisting of a sensor and a data processor enabled for algorithmic calculation. The latter provides interpretation or perception services based on the sensor data. This is often also referred to as *intelligent sensor*. The sensor modules used in this work can be grouped into the classes range sensors, imaging sensors and proprioceptive sensors. Their functional principle will be described in the following.

Range Sensors

Range sensors apply an active measurement principle to determine the distance to an object. Object positions in polar coordinates are obtainable if a directional reference is available.

Radar (radio detection and ranging) sensors use electro-magnetic waves for distance measurement. Radar waves are transmitted into a designated direction in space by a directive antenna. The waves are reflected from any object representing a large change in the dielectric constant or diamagnetic constant. If the wavelength is much shorter than the target's size, the wave will bounce off in a way similar to the way light is reflected by a mirror. A receiver antenna located at the radar module can detect the reflected wave.

An object reflects a limited amount of radar energy. Objects with conducting surfaces, such as metal, reflect strongly, so that objects like cars can be detected very well. Objects like pedestrians can be detected in principle but entail much lower reflection amplitudes due to higher absorption. The RCS is a measure for the detectability of an object. The RCS of a radar target is the hypothetical area required to intercept the transmitted power density at the target such that if the total intercepted power was re-radiated isotropically, the power density actually observed at the receiver is produced [10]. A target's RCS depends on its relative size to the wavelength, the reflectivity of its surface, the direction of the radar reflection caused by the target's geometric shape, and the polarization with respect to the target's orientation.

Two classes of radar sensors can be distinguished: continuous wave (CW) radars and pulse Doppler radars. A so-called FMCW radar (Frequency Modulated Continuous Wave) results if the frequency of the transmitted wave of a CW radar is periodically modified (sweep). The frequency shift between the reflected wave and currently sent waves enables the computation of the signal travel time τ_t. The multiples of the sweep period time lead to an ambiguous determination of the radial distance r to the object using the speed of light c:

$$r = \frac{c\tau_t}{2}.$$

(2.1)

Pulse Doppler radars transmit short wave packages in certain time intervals. The signal travel time τ_t between transmission and reception is utilized for distance computation (see Equation 2.1). Radar sensors can detect the velocity of objects relative to the propagation direction of the wave using the Doppler effect[1] (range rate $\dot{r}$). The maximum unambiguously detectable relative speed depends upon the cycle duration of a sweep.

There are three typical methods to obtain angle measurements as directional reference of the range sensor. *Manual scanning radars* use mechanical deflection of a concentrated beam to measure their environment in multiple steps. *Multiple beam radars* use the superposition of several transmission beams with different directions. The known receiver characteristic enables the inference about the angle with the strongest reflection. *Single beam radars* require several receiver antennas at different positions. The phase difference of a reflected wave between the single receiving antennas can be utilized to compute the angle of a detected object.

The utilized automotive radars work at frequencies around 76 GHz to 77 GHz. Therefore, objects in ranges at which other electro-magnetic wavelengths, such as visible light or infrared light, are attenuated too much, can be detected by radar since the radio waves are barely absorbed by the medium through which they pass. This is beneficial, as radar sensors can be designed in a way that they also work during bad weather conditions like fog, rain or snow fall. On the one hand, radar sensors are well suited for the separation of stationary and moving objects due to their capability of direct, radial speed measurement. On the other hand, the discrete and widening beams limit the angular resolution. Moreover, multi-path propagation and diffraction can lead to false measurements, i.e., ghost objects. Measurements become inaccurate if a part of the radiation diffuses and does not reflect at an object's surface. However, currently applied automotive radar sensors provide radial distance measurements with standard deviations below 10 cm [11].

Lidar sensors (light detection and ranging) utilize the travel time of concentrated, monochromatic light for distance measurement. A transmitting diode creates a laser (light amplification by stimulated emission of radiation) pulse that reaches the surface of an object where optic effects like refraction and reflection are induced. The reflected part of the light energy can be detected by a receiving diode. The relative velocity using the Doppler effect is not determined in automotive applications due the enormous metrological effort that would be required for determination of the small frequency difference between transmitted and received signals. Angular measurements are performed using an arrangement of several

---

[1] The Doppler frequency is computed by using the frequencies of the transmitted and the received signal.

directed diodes (multi-beam-lidar) or using a rotating mirror that deflects the laser beam to several different directions. Beside lidars providing information about distance and horizontal angle of an object, there are laser scanners that additionally measure vertical angles of the reflections.

On the one hand, distance measurement with lidar sensors is very accurate ($<$ 2 cm [12]) since the beam is usually reflected at the surface of the object. Moreover, the angular resolution, especially of laser scanners, is very high (up to $0.09°$ [12]). On the other hand, the optical measurement principle can lead to erroneous measurements in rain, snow and fog or when the sensor is spoiled. Furthermore, laser scanners are much more expensive than the automotive sensors described above.

Imaging Sensors

Monocular cameras (grayscale or color) represent the environment as 2D image. Color cameras usually provide a lower resolution than grayscale cameras due to the required measurement channels for color separation, but they enable the separation of different color spectra. A color filter array covers the photo sensor using 50 % green, 25 % red and 25 % blue which results in a so-called Bayer pattern.

Today's automotive cameras use CMOS (complementary metal oxide semiconductor) sensors for photon detection that exploit the photo effect (imager). Compared to charge-coupled devices (CCD), CMOS sensors do not suffer from the blooming effect where a light source overloads the sensitivity of the sensor causing the sensor to bleed the light source onto other pixels. They consume less power than CCDs and require less specialized manufacturing facilities rendering them less expensive. Automatic exposure control (shutter) and contrast correction can be realized dependent on the application and outer illumination and can thereby reduce motion blur.

Automotive computer vision has to handle highly dynamic environments such as low sun and shadow. The available bit depth per pixel is typically limited (i.e. 8 bit), so that the exposure control is optimized for a certain region of interest (ROI). High dynamic range imaging (HDRI) techniques combine several image frames with different exposure times using non-linear coding of the image data and thereby expand the dynamic range of the images. Regarding the choice of the camera's lens, it is important to consider the focal length f_{opt} and the maximum aperture angle. The focal length determines the magnification of objects projected onto the image plane, and the aperture angle is responsible for the light intensity of an image. Shorter focal lengths give a wider FOV compared to longer focal length lenses. The wider the aperture angle, the faster the shutter speed may be for the same exposure. Wide-angle lenses ($60°$ to $114°$ horizontal aperture) suffer from low resolution and low quality at long distances, while normal lenses ($28°$ to $60°$ horizontal aperture) balance between a wide FOV and a high resolution at long distances. Typical automotive cameras provide 25 to 80 frames per second (FPS). On the one hand, a high angular resolution and a high information content foster a high-quality analysis of the image by classifiers. On the other hand, this induces a high data rate that needs to be processed in real-time, so that a tradeoff is made between the aperture angle, the image resolution, and the computational demand. The texture strongly depends on illumination and weather conditions, which might influence the classification process. The three-dimensional world is projected onto the two-dimensional image, so that information is lost. Power flow[2] tries to estimate depth information from motion and can compensate for that loss in a limited way, but is computationally expensive.

Stereo cameras consist of an arrangement of two monocular cameras with a known distance (basis). Disparities between corresponding pixel pairs of both cameras are utilized for distance measurement of an detected object. Finding pixel correspondences is computationally demanding and ambiguous, and wrong assignments, especially in periodic textures, lead to incorrect distance computations. Furthermore, the distance accuracy is limited by the resolution, especially at long distances.

Proprioceptive Sensors

Proprioceptive sensors measure signals originating from within the vehicle. The sensors monitor the internal status and serve the determination of acceleration, speed, orientation and relative position of the vehicle. GPS (global positioning system) sensors, encoders and gyroscopes are examples for proprioceptive sensors. Typical accelerometers and gyroscopes in automotive applications are fabricated as micro-electro-mechanical systems (MEMS). Here, the sensors send their data via CAN to the processing computer. CAN is a multi-master broadcast serial bus standard for connecting ECUs. In case that several nodes try to send messages at the same time, the message with the highest priority (smallest nu-

---

[2]Power flow is caused by relative motion between object and camera.

merical ID value) overwrites all other messages. The data is used to gain information about the ego vehicle's state and to predict its future behavior. More detailed information about accelerometers, gyroscopes, wheel speed encoders and the measurement principle of the steering wheel angle can be found in Appendix A.1.

2.2.2 Sensor Data Fusion

Data fusion is a formal framework comprising means and tools for the alliance of data originating from different sources [13]. Its general purpose is to obtain information of enhanced quality compared to information from single sensor solutions where the exact definition of 'enhanced quality' is application-specific. Data fusion is a multi-disciplinary research area borrowing ideas from many diverse fields, such as signal processing, information theory, statistical estimation and inference, and artificial intelligence. Information fusion is an integral part of the perception of numerous technical and biological systems [14]. The core advantages of a multi-sensor system are an enhanced detection certainty and measurement accuracy when using redundant data as well as an enlarged FOV and an increased number of state variables that can be measured directly when using complementary sensors. *Accuracy* is defined as the mean deviation of the state estimation from the reality, and it is influenced by the sensors and the applied algorithms. The data fusion approach shall handle uncertain, incomplete, or even defective data. Outliers and spurious data should be identified using the redundancy which enhances the confidence and reliability of measurements. *Reliability* is referred to as the correctness of a statement in relation to an object existence. Moreover, the estimation result has to be consistent.[3] Extended spatial and temporal coverage improves the detection performance. A good detection performance comprises a high true positive rate compared to the false alarm rate and it is often described by receiver operating characteristic (ROC) curves.

A well-designed data fusion method should incorporate multiple time scales in order to deal with timing variations in data (data rates). Incorporation of a recent history of measurements into the fusion process is desirable in dynamic environments to obtain information about dynamic processes that would not be observable otherwise. Heterogeneous and

---

[3] Interpretation in statistics: Increasing the sample size has the effect that an estimator's result is getting closer to the real value.

homogeneous data should be processable (data modality). Data alignment or registration takes care of the transformation from each sensor's local frame into a common frame before fusion.

The vehicle's environment model shall be devised in such way that it is suitable for different applications at once. Thus, if an additional application is added that requires a lower or higher confidence about the objects in the environment than an existing application, there should be no need for a change in the fusion algorithm or in the interface. For example, the different applications should be able to react based on objects having a sufficient PoE.

Numerous simplifications are required regarding the level of detail of the underlying model classes as well as their number and type diversity due to the complexity of the real world.

There are different possibilities to categorize the structures and methods of data fusion approaches. Distinctions can be made by the architecture, the abstraction level of the input data, or the sensor integration. Furthermore, one can distinguish between implicit and explicit fusion approaches as well as between grid-based and parametric approaches. These categories will be described in the following.

Durrant-Whyte [15] classifies a sensor system into three basic **sensor types** where all three classes may be present simultaneously in real-world applications:

- *complementary* (supplementary): combine incomplete and independent sensor data to create a more complete model, e.g., by utilizing different FOVs due to distinct mounting positions or different measurement principles,

- *competitive* (redundant or contrary measurement results): reduce effects of noisy and erroneous measurements, increase reliability, accuracy and decrease conflicts where contrary events require special decision processes,

- *cooperative* (enhancing the quality): extension of the measurement space by one stochastic state-space dimension that cannot be measured directly (higher-level measurement).

The sensor data can be fused implicitly or explicitly. **Implicit fusion** processes the measurements and updates the model (temporal filtering) when new measurements are available (non-deterministic), which requires consistent data processing. Data can be associated based on a sensor-specific abstraction level, which is advantageous when using heterogeneous sensors that measure different state dimensions. Furthermore, asynchronous sensors are applicable and adding additional sensors is relatively simple, as they are not coupled in the association and filter update step. Last but not least, the implementation of a fault-tolerant system behavior is usually simpler than for explicit fusion. In contrast, **explicit fusion** fuses all data sources in one association step at the same time requiring temporal filtering and a common abstraction level of the data processing for all sensors (dependent on the object description). The requirement of synchronous sensors or measurements is often hard to fulfill with standard automotive sensors.

A synchronous and thus deterministic system behavior is desirable, but not all of today's automotive sensors can be triggered, so that the system has to be designed as an asynchronous system due to technical limitations.

The **abstraction level** at which the data fusion takes place has to be chosen carefully and it is a trade-off between information content and complexity. Three different abstraction levels can be distinguished according to [16]:

- *Signal / raw data level*: direct combination of signals from different sensors which requires comparability of the measured signals as well as their registration and synchronization; highest information content as there is no data reduction causing an immense data volume,

- *Feature level*: fusion of signal descriptors to obtain enhanced estimates of certain signal characteristics is useful if there is no spatial and temporal coherence of the data,

- *Symbol level*: combination of symbolic signal descriptors (symbols, objects, decisions, equivalence classes) to make decisions based on associated probabilities.

Figure 2.2 illustrates the enumerated fusion levels including the suitable data models, the aim of the specific fusion and an example.

Considering the **fusion architecture**, one can differ between *centralized fusion* with one decision center, *distributed fusion* with local decision centers and a *combination* of both. Usually, limitations in communication bandwidth, required real-time capability, and the complexity of data processing lead to a mixed architecture in which information is processed

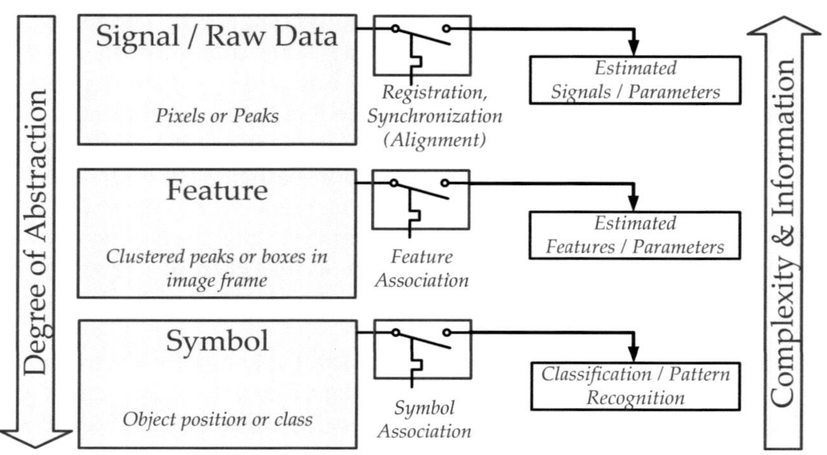

Figure 2.2 Comparison of fusion levels with types of fused data, prerequisites, suitable data models and goal.

locally in distributed decision centers, and only the processing results are transferred to a central decision center.

Finally, there are two fundamental distinctions in **fusion methods** [16]: *grid-based* (geometric) and *parametric* (numeric) approaches.
Grid-based fusion approaches combine synchronous and sequential distance measurements from homogeneous or heterogeneous sensors to obtain a so-called occupancy grid with associated obstacle probabilities (situation picture). The occupancy probabilities of the different grid locations are accumulated and compared with a threshold during the mapping process. Continual measurements from different positions update the grid. The object probability for a certain location is assumed to be high if repeated measurements indicate high occupancy values for the corresponding cells [17].
The parametric fusion approaches can be categorized into *feature-based* and *probabilistic* approaches, *fuzzy* methods and *neural* approaches. Feature-based approaches include the weighted average, the Kalman filter [18] methods and its extensions to non-linear systems (Extended Kalman filter (EKF) [19], Unscented Kalman filter (UKF) [20–22], information filter [23, 24]), while probabilistic approaches deal with meth-

ods related to classic and Bayesian statistics [25] as well as Dempster's and Shafer's theory of evidence [14, 26]. In contrast to classical statistics (maximum-likelihood (ML) frameworks), Bayesian statistics (maximum-a-posteriori (MAP) frameworks) do not only consider the observations but also the measurand as a realization of a random variable. The essential advantage of the latter is that it yields the probability distribution of the parameter of interest given the measurement data, whereas the classical approach only describes the probability distribution of the sensor data for a given parameter value [27].

Grid-based models, particle filters and Kalman filtering have been the most common data fusion techniques for a long time. Interval calculus, fuzzy logic and evidential reasoning have been proposed to deal with perceived limitations in probabilistic methods such as complexity, inconsistency, precision of models and uncertainty about uncertainty [28]. Hybrid methods combine several fusion approaches to develop a meta-fusion algorithm.

Artificial neural networks (ANN) are based on interconnected processing units to solve a specific task. These networks use supervised or unsupervised (Kohonen maps) learning mechanisms for classification and recognition and are widely used for data fusion of several complementary sensors [29]. Their application is especially advantageous if it is difficult or impossible to specify an explicit algorithm for data combination. Unfortunately, ANN lack of possibilities to incorporate prior knowledge on the magnitudes involved in the fusion task [16]. Bayesian fusion allows relatively simple descriptions of the magnitudes involved in the data fusion and therefore, it is well suited in context with measurement data [16]. Furthermore, objective knowledge and vague information can be incorporated into the fusion approach. Probability density functions (PDF) express the available knowledge in Bayesian fusion theory. Prior knowledge about the measurand is represented by the prior PDF, whereas information regarding the measurement method is described by the likelihood function. A physical model of the measurement process or an empiric determination can provide this knowledge [25].

In this work, the parametric Bayesian fusion-based approach EKF is used in a mixed architecture to fuse pedestrian data from the heterogeneous sensor types radar and camera mounted at different positions of a commercial vehicle. Data is fused on the feature level. Different methods for the association of object and measurement data are used. The methods are introduced together with the required filter equations in Chapter 4. The focus of this work is on the probabilistic modeling and not on the state estimation, so that no other filters for state estimation have been investigated. A more detailed description of the used fusion concept, the sensors and the chosen architecture follows in Chapter 4 as well.

2.3 Knowledge Base for Situation Evaluation

The sensor data fusion module provides a model of the environment by perception of its elements in time and space. Not only the tracking of obstacles is a challenge, but also obtaining detailed information about a traffic scenario including the actions of the ego vehicle and the interactions with other traffic participants.

A human driver judges the risk of a situation in advance. The attentive human is still the best known 'predictor of traffic scenarios' due to his experience and his high-resolution environment perception. However, human drivers can suffer from fatigue or might be overstrained, for instance, in situations where they should decide if people walking on the sidewalk towards the street will stop or keep on walking to cross the street. In such situations of uncertainty, drivers usually react more cautiously, e.g., drive more slowly or steer an evasion trajectory. If ADAS forced a more cautious way of driving in such situations due to incomplete situation awareness, this would not be accepted by most drivers. Hence, an ADAS has to reach a sufficient level of certainty that the situation being judged is risky before it intervenes. The guideline ISO31000 [30] describes risk as the combination of the *occurrence probability of a harmful event* with the amount of *damage caused by the event*.

The situation evaluation module interprets the meaning of situation elements and forecasts their states for the near future. The aggregation of all relevant information serves the understanding of the situation, which is the basis for making a correct decision regarding the allocated objectives or the desired end-state. The chosen action should then change the situation in the most preferable way.

The environment model of a car is much less complete and includes less details than the environment model of a human, so that only a limited number of situations can be recognized and differed. Therefore, situations are divided into classes. A recognized situation is assigned to the class which it is most similar to (situation mapping) [31]. ADAS are developed for the reaction in very specific situations. A typical situation would be 'risk of rear-end collision' and may include the following available patterns: a car is in front on the ego lane, the car in front decelerates, the distance between the own car is below a speed-dependent threshold value. Then, there could be three possible reactions to this situation: brake to avoid the collision, evade the car by steering or keep-on-driving and collide with the car in front. Each reaction implies a risk since braking might evoke a collision with the following traffic and steering might induce a crash with the oncoming traffic or the road border or might lead the vehicle off the road. Since only a limited number of patterns and details can be captured, it is not possible to evaluate every hazard, to create more patterns and to differ between more sub-situations. Therefore, the possible reactions are usually limited to braking or keep-on-driving in this situation. The task of the situation evaluation module in the driver assistance system would be to recognize — based on the available patterns — if a situation is as critical that a specific reaction should be applied or not.

Information about detected elements in the environment and about the ego vehicle's state itself underlies spatial and temporal inaccuracies. Furthermore, there might be objects in the environment that have not been detected or even false detections. Thus, the situation evaluation is based on a more or less inaccurate, incomplete and uncertain environment model. Moreover, two situations may develop in different ways, although the setting initially looked the same. In a situation where a pedestrian approaches the lane from the right, the pedestrian might either stop at the road edge or might keep on walking onto the lane, so that a collision with the own vehicle could follow. There are uncertainties in the prognosis. All these facts have to be taken into account implicitly (in the general system design) or explicitly (in each situation). A sufficient level of certainty compared to the induced hazard has to be reached before a system reaction is applied.

A prediction model is required to forecast the evolution of the situation. Constant velocity (CV) models are commonly used to predict the future states of pedestrians, while constant turn models in combination with constant speed (CTCV) or constant acceleration (CTCA) models are mostly used to forecast the future states of vehicles. Modern approaches do not only consider a linear evolution, but include variations of speed, acceleration and yaw rate within certain boundaries.

ADAS are designed for specific use cases that are deviated from detectable, classifiable, and distinguishable situations that often lead to accidents according to accident analysis and statistics. On the one hand, these accident statistics help to identify situations that often lead to an accident as well as implicated hazards. An adequate system reaction in these situations is most effective for the reduction of the number of traffic victims. On the other hand, accident statistics illustrate how situations evolve and thus, they support the development of models for situation prognosis.

Inattention of the driver during the last three seconds before the collision is a contributing factor in 93 % of the crashes with passenger cars [32] where inattention can be categorized into secondary task engagement, fatigue, non-specific eye glance, and driving-related inattention (i.e., due to checking of the rear-view mirrors or the blind spot). Thus, many accidents could be avoided or reduced in severity if the driver gets a warning in this time frame.

The next subsections describe the statistical relevance of situations that often lead to accidents involving pedestrians. The following subsection concentrates on typical pedestrian behavior, which is an important basis to forecast pedestrian states, to predict upcoming situations, and to estimate the risk of a setting.

2.3.1 Studies on Pedestrian- and Truck-related Accidents

The development of ADAS avoiding potential accidents with vulnerable road users requires a thorough understanding of how the settings of these accident types look like and how often they take place. Collisions between vehicles and pedestrians or bicyclists are evaluated as particularly critical since vulnerable road users are unprotected and often suffer from lethal injuries as a result of crashes. Where people live can also influence their exposure to road traffic risk. In general, people living in urban areas are at greater risk of being involved in road crashes, but people living in rural areas are more likely to be killed or seriously injured if they are involved in crashes. One reason is that motor vehicles tend to travel faster in rural areas [3].

Traffic strongly differs between Western countries and countries with emerging markets like India or China, especially in the urban areas. This work refers to the development of ADAS for countries with traffic comparable to that in Germany, the USA or Japan.

In Germany, most accident studies are based on the database GIDAS (German In-Depth Accident Study), the database of the German Federal Statistical Office and the database of the German Insurers Accident Research (IAR). The information depth of the IAR database increases the one of the Federal Statistical Office, but is limited regarding the informative value of some accident database attributes compared to GIDAS [33, 34] since there has been no analysis directly at the site of the accident [35]. Only a few collisions of commercial vehicles with vulnerable road users are described in detail. More information is available regarding accidents involving passenger cars and vulnerable road users. It is expected that most of the settings that lead to a collision of a vulnerable road user look similar for passenger cars and commercial vehicles. Moreover, most aspects of the developed approach for risk assessment can be transferred to passenger cars. Therefore, statistics for both classes are presented.

About 14 % of all passenger car accidents and about 20 % of all passenger car primary collisions with killed or seriously injured persons in the IAR database included pedestrians. Hummel et al. [35] found that pedestrians are recognized as critical very late and that more than 70 % of collisions between passenger cars and pedestrians in the database took place at car speeds of about 30 km/h.

The GIDAS database of 2009 and the database of the German Federal Statistical Office of 2009 provides the information that 7.7 % of all accidents involving a commercial vehicle and another road user have been collisions with pedestrians, whereas the fraction of killed persons in these accidents is even higher: 10.4 %. 90 % of the pedestrian accidents that include a commercial vehicle take place in the built-up environment.

The DEKRA report of 2011 [2] states that two thirds of all accidents in Europe take place within built-up areas where 47 % of the killed road users are pedestrians or bicyclists. These vulnerable road users are often insufficiently lit during twilight and during the night. They are often recognized too late due to several reasons like occlusion by other objects or driver distraction. Pedestrians older than 65 years and children are particularly endangered.

Half of the vulnerable road users that were killed in German traffic in 2009 have been **older than 65 years**, although people older than 65 years represented only about 20 % of the population. Older pedestrians have problems to overlook traffic from various directions at the same time when they cross wide streets or intersections with dense traffic [36]. Age-related physical, perceptual and cognitive limitations and the missing ability to compensate for these limitations make it difficult for older people to perceive if a gap is sufficient for them to cross the street [37]. Additionally, younger road users tend to underestimate the accident risk for older road users [38]. The walking speed that is considered as normal is not reached by about a tenth of the older persons [39] and it takes more time for older people to perceive and react upon light signals. Older bicyclists often do not watch out for traffic from behind and have problems

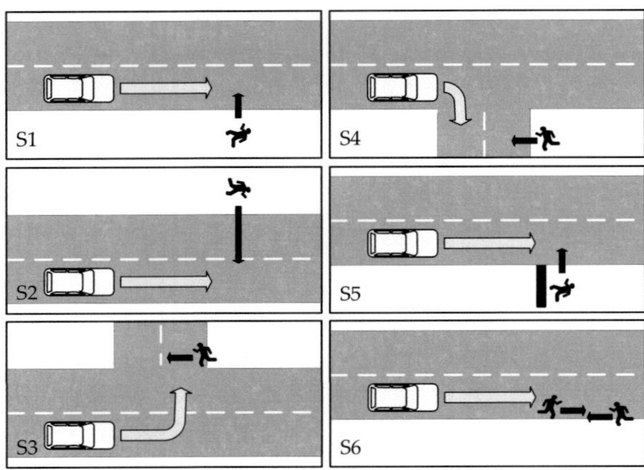

Figure 2.3 Situations with passenger cars and pedestrians that frequently lead to accidents. Adapted from [41].

maneuvering the bicycle [40]. Moreover, Bernhoft et al. [40] state that although the older road users generally are more careful than the younger, it seems that health problems may cause older persons to reconsider their behavior in favor of a more risky one, e.g., cross the road irrespective of crossing facilities nearby in order to avoid detours.

Children younger than 15 years constitute the second group that is particularly endangered when moving in traffic as pedestrian or bicyclist [2]. Firstly, they are less experienced. Secondly, they are smaller than older pedestrians, so that cars or other objects occlude them more easily.

Typical situations leading to collisions with pedestrians have to be identified to develop an ADAS for the avoidance of accidents. The action group vFSS (German: vorausschauende Frontschutzsysteme) [41] has analyzed the GIDAS database to identify typical accident situations of pedestrians with passenger cars that occur in Germany and abroad. Accidents involving passenger cars and vulnerable road users lead the statistics, followed by those involving commercial vehicles. About 60 % of the collisions where road users were injured or killed in built-up areas took place at crosswalks (42.2 %), at stops (12.7 %) or in restricted traffic zones (5.7 %) [2]. These are regions where drivers could expect pedestrians, but often they do not see them.

Situations where pedestrians cross the streets from different sides with and without occlusion (S1, S2, S5) have been documented most frequently (see Figure 2.3). Turning maneuvers of passenger cars (S3, S4) lead less frequently to accidents. Collisions where the pedestrians walk along the road typically occur on country roads and highways (S6). Situations with pedestrians crossing the street led to 817 accidents with injured persons (2.3 %) and to a fraction of 5.6 % of the accidents in which people got killed [41].

The 50 %-random sample of the German Federal Statistics Office (years 2005 to 2009) includes 84116 commercial vehicles with a maximum permissible weight of 6 tons and more. 49 % of the collisions with a pedestrian occurred during crossing situations.

The analysis of Hummel et al. [35] showed that about 13 % of the accidents with commercial vehicles occur during turning maneuvers. A vehicle primarily collided with a pedestrian or a bicyclist in 80 % of these cases. It is assumed that a system that warns the driver from obstacles in the vehicle's blind spot at the right side before and during turning maneuvers could have avoided 31.4 % of the lethal turning accidents and 43.5 % of those ones with seriously injured persons. This means that such a system could have avoided 4 % of the lethal accidents and 5 % of all accidents with seriously injured persons related to the total number of accidents that include commercial vehicles.

2.3.2 Pedestrian Motion Analysis

A pre-requisite for the successful development of a system that avoids collisions with pedestrians is the reliable detection and tracking of pedestrians. But that is not enough to achieve a good balance between reactions to pedestrians that are a hazard and false positive reactions, e.g., due to pedestrians at the pavement who may not intend to cross the street. Hence, influencing factors and parameters have to be identified that allow an unambiguous prediction of the intention and behavior of a pedestrian. The trajectory is only one possible parameter and subject to errors since there is no trajectory estimation for a stationary pedestrian. Furthermore, pedestrians can start walking suddenly, change their direction abruptly or stop. A pedestrian walking parallel to the street might not be considered as a hazard by a collision avoidance system that is only based on trajectory computation although the pedestrian might continuously turn his head towards traffic and could cross the street in less than one second.

Himanen and Kumala analyzed 799 events with pedestrians and vehicles. Their results indicate that the most important explanatory variables to predict the probability that a pedestrian crosses the street were the pedestrian's distance from the kerb, the city size, the number of pedestrians crossing the street simultaneously, the vehicle speed and the vehicle platoon size [42].

The time gap until car and pedestrian would collide (time-to-collision — TTC) as an additional parameter has been analyzed in [43]. The authors found a minimum time gap size of 2 seconds and a maximum of 11 seconds. On roads with smaller width, tighter gaps were chosen. A smaller TTC range of 3 to 7 seconds has been identified in a study with German pedestrians in [44]. Similarly to Oxley et al. [37], results propose that distance information, rather than the TTC, is the relevant criterion on the basis of which pedestrians decide to cross the street or not. When cars approached with higher speeds, the test participants chose larger gaps but smaller TTCs. Consequently, a driver has to expect a more dangerous behavior from pedestrians when he is driving faster.

Pedestrian motion parameters are important to design a system that avoids collisions with pedestrians. Several user and environmental factors influence pedestrian speeds. Pedestrian speed and acceleration parameters have been analyzed in several studies dependent on gender, age, state of the traffic light signal, road geometry and social behavior (single pedestrians or groups) [45–48]. Table A.1 in Appendix A.2 summarizes the results of the corresponding authors. According to these authors, the mean speed of pedestrians is about 1.4 m/s. Older pedestrians walk more slowly so that their mean speed of 0.9 m/s lies below the average of all pedestrians. Moreover, jogging pedestrians reach speeds of about 3 m/s, while running male pedestrians reached a mean speed of 5 m/s.

Crosswalks with traffic lights are designed for a pedestrian speed of 1.22 m/s (4 ft/s) according to [47]. Carey [45] found that pedestrians starting their crossing during the flashing *DON'T WALK* sign typically cross faster than those who start during the *WALK* sign. This is plausible since these pedestrians know that they are running out of time. Furthermore, the road users moved slightly slower when crossing with others instead of alone.

3 Theoretical Foundations and Methods

This chapter describes the methods that have been used in this thesis and outlines the required theoretical foundations. Sensor data is fused to detect and track objects in the environment. The required steps are illuminated in the first two sections. Classification techniques are presented in Section 3.3. They enable the detection of pedestrians in images and help to distinguish between different maneuver classes of the ego vehicle. Stochastic reachable sets are computed for the ego vehicle and the detected road users to predict their future states. Therefore, the last sections are dedicated to the description of discrete event systems and hybrid automatons, reachability analysis and Markov chain abstraction.

3.1 Object Tracking

There are several approaches for object tracking, as described in the previous chapter. This work focuses on Bayesian state estimation, the Kalman filter and its extension to non-linear systems (EKF). The applied feature-based object tracking process can be divided into the following steps, as depicted in Figure 3.1.

Raw data is recorded by the sensors in the first step. The second step provides *pre-filtered data* that is referred to as measurements in the following. Several approaches exist for *association* of tracked objects with measurements and will be described later. The *update* of the objects' state estimations is the central element of the tracking procedure and it is described in the following subsections. The *object management* module decides whether objects should be deleted in case there are no more measurements available for that object or when the object's PoE is very low. Furthermore, it decides which measurements should lead to the creation of new objects in the environment model (object birth).

3.1.1 Bayesian State Estimation

Object tracking algorithms try to estimate the number of objects in a scene and they compute the objects' states over time utilizing uncertain measurement data $\mathbf{z}_k$ (imprecise, noisy, defective or missing) from sensors

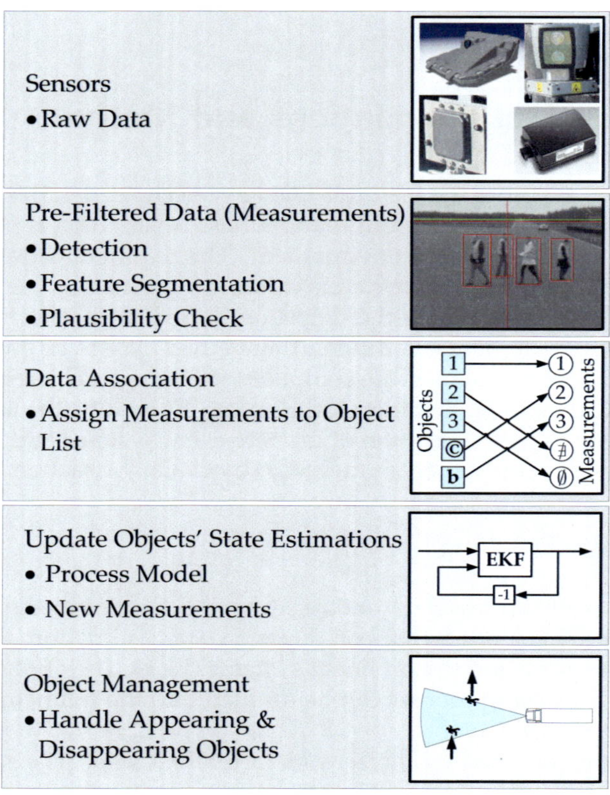

Figure 3.1 Steps of the Object Tracking Procedure.

and prior knowledge about the sensor properties (measurement principle) $p(\mathbf{Z}|\mathbf{X})$, where $\mathbf{X}$ are all object states. Moreover, dynamic systems provide information about the state's history and about the object's behavior over time (system dynamics).

Applying the Bayesian rule [49, 50] with pre-existing knowledge of all measurements $\mathbf{Z}_{1:k} = (\mathbf{z}_1, \ldots, \mathbf{z}_k)^{\mathrm{T}}$ until time point k, one can compute the knowledge base at time point k:

$$
\begin{aligned}
p(\mathbf{x}_k|\mathbf{Z}_{1:k}) &= \frac{p(\mathbf{Z}_{1:k}|\mathbf{x}_k)p(\mathbf{x}_k)}{p(\mathbf{Z}_{1:k})} = \frac{p(\mathbf{z}_k,\mathbf{Z}_{1:k-1}|\mathbf{x}_k)p(\mathbf{x}_k)}{p(\mathbf{z}_k,\mathbf{Z}_{1:k-1})} & (3.1)\\[2mm]
&= \frac{p(\mathbf{z}_k|\mathbf{Z}_{1:k-1},\mathbf{x}_k)p(\mathbf{Z}_{1:k-1}|\mathbf{x}_k)p(\mathbf{x}_k)}{p(\mathbf{z}_k|\mathbf{Z}_{1:k-1})p(\mathbf{Z}_{1:k-1})} & (3.2)\\[2mm]
&= \frac{p(\mathbf{z}_k|\mathbf{Z}_{1:k-1},\mathbf{x}_k)p(\mathbf{x}_k|\mathbf{Z}_{1:k-1})}{p(\mathbf{z}_k|\mathbf{Z}_{1:k-1})}. & (3.3)
\end{aligned}
$$

The PDF of the measurement principle in Equation 3.3 depends on all preceding measurements until time point k. It is assumed that the current measurement depends only on the current state and not on the state history or measurement history ($p(\mathbf{z}_k|\mathbf{X}_{1:k},\mathbf{Z}_{k:k-1}) = p(\mathbf{z}_k|\mathbf{x}_k)$). Thus, Equation 3.3 can be simplified according to:

$$
\begin{aligned}
p(\mathbf{x}_k|\mathbf{Z}_{1:k}) &= \frac{p(\mathbf{z}_k|\mathbf{x}_k)p(\mathbf{x}_k|\mathbf{Z}_{1:k-1})}{p(\mathbf{z}_k|\mathbf{Z}_{1:k-1})} & (3.4)\\[2mm]
&= \frac{p(\mathbf{z}_k|\mathbf{x}_k)p(\mathbf{x}_k|\mathbf{Z}_{1:k-1})}{\int p(\mathbf{z}_k|\mathbf{x}_k)p(\mathbf{x}_k|\mathbf{Z}_{1:k-1})\mathrm{d}\mathbf{x}_k}. & (3.5)
\end{aligned}
$$

The size of the measurement history $\mathbf{Z}_{1:k}$ and the state history $\mathbf{X}_{1:k}$ are growing linearly over time, so that they exceed all memory capacities. Therefore, technically realizable measurement systems can only observe dynamic systems that incorporate the Markov property [51]: The Markov process of first order states that the current state depends only on the previous state. It can be applied without loss of generality, and thus:

$$
p(\mathbf{x}_k|\mathbf{Z}_{1:k-1},\mathbf{X}_{1:k-1}) = p(\mathbf{x}_k|\mathbf{x}_{k-1}). \tag{3.6}
$$

The prior knowledge base $p(\mathbf{x}_k|\mathbf{Z}_{1:k-1})$ can now be computed recursively based on the knowledge base of the previous time step using the Chapman-Kolmogorov equation:

$$
p(\mathbf{x}_k|\mathbf{Z}_{1:k-1}) = \int_{\mathcal{X}} p(\mathbf{x}_k|\mathbf{x}_{k-1})p(\mathbf{x}_{k-1}|\mathbf{Z}_{1:k-1})\mathrm{d}\mathbf{x}_{k-1}. \tag{3.7}
$$

Process or motion models provide the transition probability $p(\mathbf{x}_k|\mathbf{x}_{k-1})$ and make it possible to incorporate problem-specific knowledge about motion limitations, directions, or object information. They enable the estimation of the object state at time point k. Their description contains a (time-variant) function f including stochastic noise $\mathbf{v}_k \sim \mathcal{N}(\cdot,\mathbf{0},\mathbf{Q}_k)$ with covariance $\mathbf{Q}_k$ to account for the model error due to uncertainty about the exact model motion or due to approximation:

$$\mathbf{x}_k = f(k, \mathbf{x}_{k-1}, \boldsymbol{\nu}_k).$$ (3.8)

Eventually, an external, deterministic controller input $\mathbf{u}_k$ can influence the system using the controller function g:

$$\mathbf{x}_k = f(k, \mathbf{x}_{k-1}, \boldsymbol{\nu}_k) + g(\mathbf{u}_k).$$ (3.9)

Measurements from different sensors help to improve the state estimation of an existing object due to their different positive properties. The measurements are included in an update step. The measurement model of the sensor provides the likelihood function of the measurement $p(\mathbf{z}_k|\mathbf{x}_k)$, describes the measurement process of the sensor and is represented by the measurement function h. It transforms the state space to the sensor-dependent measurement space and enables that the state update (innovation) is performed directly based on the sensor data:

$$\hat{\mathbf{z}}_k = h(k, \mathbf{x}_k, \boldsymbol{\eta}_k),$$ (3.10)

where $\boldsymbol{\eta}_k \sim \mathcal{N}(\cdot, \mathbf{0}, \mathbf{R}_k)$ with covariance $\mathbf{R}_k$ represents stochastic noise that accumulates measurement errors.

In conclusion, the recursive Bayesian state estimation is accomplished in two phases. First, the knowledge base is projected onto the next time point of a measurement in the prediction step using the process model and Equation 3.7. Next, information from the current measurement is included into the knowledge base in a correction step (update or innovation) using the Bayesian rule, see Equation 3.5.

Subsections 3.1.2 and 3.1.3 introduce two concrete implementations of the Bayesian state estimation for one object using linear and non-linear functions. Subsection 3.1.4 provides the required extensions for multi-object tracking.

3.1.2 Kalman Filter

The Kalman filter [18] recursively estimates the state of a time-variant state variable with minimum error variance in the presence of noisy data in a time-discrete system. It is designed for Gaussian distributed state variables. Linear functions enable a simplified representation of the process model (Equation 3.8) and the measurement model (Equation 3.10) using matrix multiplications:

$$\mathbf{x}_k = \mathbf{F}_k\mathbf{x}_{k-1} + \boldsymbol{\nu}_k \text{ and} \tag{3.11}$$

$$\mathbf{z}_k = \mathbf{H}_k\mathbf{x}_k + \boldsymbol{\eta}_k, \tag{3.12}$$

where $\mathbf{F}_k$ represents the system matrix (process model matrix) and $\mathbf{H}_k$ the measurement matrix at a time point k. The noise terms $\boldsymbol{\nu}_k$ and $\boldsymbol{\eta}_k$ have to result from a zero-mean, white Gaussian noise process ($E\{\boldsymbol{\nu}_k\} = E\{\boldsymbol{\eta}_k\} = \mathbf{0}$). Furthermore, the requirement for their mutual, statistic independence is ascertained by evanescent auto-correlations for the time shifts $\kappa \neq 0$ (white noise) and an evanescent cross-correlation:

$$\Phi^{\nu\nu}(\kappa) = E_k\{\boldsymbol{\nu}_k\boldsymbol{\nu}_{k-\kappa}^{\mathrm{T}}\} = \mathbf{Q}\delta_{\mathrm{KD}}(k, k-\kappa), \tag{3.13}$$

$$\Phi^{\eta\eta}(\kappa) = E_k\{\boldsymbol{\eta}_k\boldsymbol{\eta}_{k-\kappa}^{\mathrm{T}}\} = \mathbf{R}\delta_{\mathrm{KD}}(k, k-\kappa), \tag{3.14}$$

$$\Phi^{\nu\eta}(\kappa) = E_k\{\boldsymbol{\nu}_k\boldsymbol{\eta}_{k-\kappa}^{\mathrm{T}}\} = \mathbf{0} \ \forall \kappa, \tag{3.15}$$

where $\delta_{\mathrm{KD}}(k, k-\kappa)$ is the Kronecker Delta function. The Kalman filter aims to compute the distribution of the knowledge base $p(\mathbf{x}_k|\mathbf{Z}_{1:k}) = \mathcal{N}(\mathbf{x}_k, \hat{\mathbf{x}}_k, \hat{\mathbf{P}}_k)$. Its estimated mean and the estimated covariance of the state are predicted using the transition probability $p(\mathbf{x}_k|\mathbf{x}_{k-1}) = \mathcal{N}(\mathbf{x}_k, \mathbf{F}_k\mathbf{x}_{k-1}, \mathbf{Q}_k)$ which results in

$$\hat{\mathbf{x}}_{k|k-1} = \mathbf{F}_k\hat{\mathbf{x}}_{k-1|k-1} \ \text{ and} \tag{3.16}$$

$$\hat{\mathbf{P}}_{k|k-1} = \mathbf{F}_k\hat{\mathbf{P}}_{k-1|k-1}\mathbf{F}_k^{\mathrm{T}} + \mathbf{Q}_k, \tag{3.17}$$

where $\mathbf{Q}_k$ is the covariance matrix of the process noise. The measurement matrix serves the transform of the predicted state and of its covariance to the measurement space:

$$\hat{\mathbf{z}}_{k|k-1} = \mathbf{H}_k\hat{\mathbf{x}}_{k|k-1}, \tag{3.18}$$

$$\hat{\mathbf{R}}_{k|k-1} = \mathbf{H}_k\hat{\mathbf{P}}_{k|k-1}\mathbf{H}_k^{\mathrm{T}}. \tag{3.19}$$

The innovation step accounts for the measurement with its state-dependent likelihood function $p(\mathbf{z}_k|\mathbf{x}_k) = \mathcal{N}(\mathbf{z}_k, \mathbf{H}_k\mathbf{x}_k, \mathbf{R}_k)$ which requires the residuum γ_k between measurement $\mathbf{z}_k$ and the measurement prediction $\hat{\mathbf{z}}_{k|k-1}$. The corresponding covariance matrix $\mathbf{S}_k$ is called innovation covariance matrix:

$$\gamma_k = \mathbf{z}_k - \hat{\mathbf{z}}_{k|k-1}, \tag{3.20}$$

$$\mathbf{S}_k = \hat{\mathbf{R}}_{k|k-1} + \mathbf{R}_k = \mathbf{H}_k\hat{\mathbf{P}}_{k|k-1}\mathbf{H}_k^{\mathrm{T}} + \mathbf{R}_k. \tag{3.21}$$

The innovation covariance matrix $\mathbf{S}_k$ and the predicted state covariance matrix $\hat{\mathbf{P}}_{k|k-1}$ serve the computation of the filter gain $\mathbf{K}_k$ (Kalman gain):

$$\mathbf{K}_k = \hat{\mathbf{P}}_{k|k-1}\mathbf{H}_k^{\mathrm{T}}\mathbf{S}_k^{-1}. \tag{3.22}$$

The Kalman gain weights the impact of the predicted state and the measurement on the state estimation at time point k in the update step:

$$\hat{\mathbf{x}}_{k|k} = \hat{\mathbf{x}}_{k|k-1} + \mathbf{K}_k\gamma_k, \tag{3.23}$$

$$\hat{\mathbf{P}}_{k|k} = \hat{\mathbf{P}}_{k|k-1} - \mathbf{K}_k\mathbf{H}_k\hat{\mathbf{P}}_{k|k-1} \tag{3.24}$$

$$= \hat{\mathbf{P}}_{k|k-1} - \mathbf{K}_k\mathbf{S}_k\mathbf{K}_k^{\mathrm{T}} \tag{3.25}$$

$$= [\mathbf{I} - \mathbf{K}_k\mathbf{H}_k]\hat{\mathbf{P}}_{k|k-1}[\mathbf{I} - \mathbf{K}_k\mathbf{H}_k]^{\mathrm{T}} + \mathbf{K}_k\mathbf{R}_k\mathbf{K}_k^{\mathrm{T}}, \tag{3.26}$$

where the last form should be chosen for a numerically more stable implementation (Joseph form) [23].

The filter equations show that certain measurements — represented by low values in the measurement covariance matrix $\mathbf{R}_k$ — increase the Kalman gain in favor of the measurements that are considered with a higher weight. On the other hand, a low uncertainty of the predicted state (low values in covariance matrix $\hat{\mathbf{P}}_{k|k-1}$) results in a higher weighting of the predicted state.

The initial state $\mathbf{x}_0$ is presumed to follow a Gaussian distribution with known expectation value $\hat{\mathbf{x}}_0$ and covariance matrix $\hat{\mathbf{P}}_0$. The gain matrix $\mathbf{K}$ is chosen in such a way that the expectation value of the mean squared error between the true state $\mathbf{x}_k^*$ and the estimation $E\{L\} = E\{\|\mathbf{x}_k^* - \mathbf{x}_{k|k}\|^2\}$ is minimized [18]. If the initial state and the noise matrix follow a Gaussian distribution, the distribution of the state estimation is Gaussian under linearized operations as well. Therefore, the Kalman filter is an optimal filter with respect to the squared error.

3.1.3 Extended Kalman Filter

If the function of the process model or the function of the measurement model or both are non-linear, the prediction and the innovation step do not result in a Gaussian distribution. Consequently, the Kalman filter cannot be applied anymore. The extended Kalman Filter (EKF) [23] linearizes the models at the operating point of the current state estimation, so that the recursive and efficient Kalman filter approach is applicable nonetheless. The Taylor series approximate non-linear process model functions $f(\mathbf{x})$ and measurement model functions $h(\mathbf{x})$, respectively.

In case of a first order EKF, the process matrix $\mathbf{F}_k$ of the Kalman filter is replaced by the Jacobian matrix of the non-linear function $f(\mathbf{x})$:

$$\mathbf{F}^{\mathrm{J}} = \left.\frac{\partial f(\mathbf{x})}{\partial \mathbf{x}}\right|_{\mathbf{x}=\hat{\mathbf{x}}_{k-1|k-1}} = \begin{pmatrix} \frac{\partial f_1}{\partial x^{(1)}} & \cdots & \frac{\partial f_1}{\partial x^{(n)}} \\ \vdots & \ddots & \vdots \\ \frac{\partial f_n}{\partial x^{(1)}} & \cdots & \frac{\partial f_n}{\partial x^{(n)}} \end{pmatrix}, \qquad (3.27)$$

where n is the dimension of the state space (number of components of $f(\mathbf{x})$). Accordingly, the measurement matrix $\mathbf{H}_k$ is replaced by the Jacobian $\mathbf{H}_k^{\mathrm{J}}$ of the non-linear measurement function $h(\mathbf{x})$ with linearization at the operating point $\hat{\mathbf{x}}_{k|k-1}$. The Jacobian matrices are then used for the projection of the state covariance matrix.

An EKF of second order would additionally require the Hessian matrix $\mathbf{F}^{\mathrm{H}} = \frac{\partial^2 f(\mathbf{x})}{\partial \mathbf{x}_m \partial \mathbf{x}_n}$ to determine the quadratic term of the Taylor approximation.

3.1.4 Multi-Object Tracking

Complex traffic scenarios with several road users require tracking of multiple objects at once. Multi-instance filters initialize and administrate separate state estimation filters for each object. Their alternatives are based on finite set statistics (FISST) [52]. Different realizations can be found in [52–57]. This work uses multi-instance filters where the state estimation is unambiguous at all times due to an object identification number (ID).

Multi-object tracking is much more complex than tracking a single object since the number of objects to be tracked is not constant and needs to be considered as a stochastic, time-variant variable N^*, the so-called cardinality. Furthermore, the number of measurements M per time step is also time-variant, and sensor-specific limitations lead to missing detections or false alarms. The number of measurements depends on the number of existing targets, which cannot be determined directly due to the sensor-induced variable number of measurements. Moreover, the association of measurements with objects is uncertain but must be presumed to be known for the state estimation in the Kalman filter. Finally, objects can occlude each other and can thereby influence their traceability.

The birth and disappearance of objects have to be modeled to estimate the cardinality N^* (*cardinality problem*). Simultaneously, a sensor-specific model needs to account for missing detections as well as for false alarms since they have a direct impact on the estimation of the object cardinality N^*.

Assigning the M sensor measurements of unknown origin to the N currently tracked objects is required. This step is called data association. Most approaches compute a binary association matrix $\mathbf{A} \in \{0,1\}^{N,M}$ with elements a_{ij}:

$$a_{ij} = 1 \Leftrightarrow x_i \leftrightarrow z_j, \tag{3.28}$$

where the symbol $\leftrightarrow$ represents the assignment of measurement z_j to object x_i. Usually, the Mahalanobis distance $d^{\mathrm{mh}}(x_i, z_j)$ is used as statistical measure to evaluate the association between measurement z_j and object x_i based on the predicted measurement $\hat{\mathbf{z}}_{i,k|k-1} = h(\hat{\mathbf{x}}_{i,k|k-1})$:

$$d^{\mathrm{mh}}(x_i, z_j) = \sqrt{(\mathbf{z}_j - \hat{\mathbf{z}}_{i,k|k-1})^{\mathrm{T}} \mathbf{S}_{ij}^{-1}(\mathbf{z}_j - \hat{\mathbf{z}}_{i,k|k-1})}, \tag{3.29}$$

where the innovation covariance matrix $\mathbf{S}_{ij}$ between object x_i and measurement z_j is used. Each measurement is associated with the object to which it has the lowest distance d^{mh} until there are no more measurements [58]. Usually, Greedy approaches [59] are applied to solve this local optimization problem. This so-called local nearest neighbor (LNN) method or nearest neighbor standard filter (NNSF) is fast but provides a suboptimal solution.

In contrast, the GNN is a maximum likelihood (ML) estimator for the data association problem [60] and provides the globally optimal solution $\mathbf{A}_{\mathrm{global}}$ by minimizing the sum of all distances:

$$\sum_{i=1}^{N} \sum_{j=1}^{M} d^{\mathrm{mh}}(x_i, z_j) \rightarrow \min. \tag{3.30}$$

The complexity of the corresponding algorithms is higher than for the LNN method. The trivial approach enumerates all possible combinations to identify the minimum (maximum of the ML-estimator):

$$N_{\mathrm{c}} = \frac{\max\{M, N\}!}{(\max\{M, N\} - \min\{M, N\})!}. \tag{3.31}$$

The runtime increases exponentially in the worst case ($M = N$) with the numbers of measurements and objects, respectively since there are $N!$ possible combinations. Stirling's formula [61] can be used to estimate the worst case runtime of the algorithm $\tau_r^{wc} \in O(N^N)$ (standard Landau notation [59]) that just enumerates all possible combinations. The solution of this global optimization problem is in the class of problems that are solvable within polynomial time (P-hardness). There are several algorithms for the computation of the GNN association matrix (e.g., the auction algorithm) [62], where the Kuhn-Munkres algorithm reaches a worst case complexity of $O(\max\{M, N\}^3)$ [63]. The Kuhn-Munkres theorem transforms the problem from an optimization problem into a combinatorial one of finding a perfect matching.

Here, a cost matrix-based implementation is chosen according to [64]. An equal number of object hypotheses and observations is assumed for simplicity of description, but the principle works similarly for differing numbers. In a distance matrix $\mathbf{D}$ with entries d_{ij}, each matrix row represents an object hypothesis x_i, $1 \leq i \leq N$, and each column represents an observation z_j, $1 \leq j \leq M$. The entries d_{ij} correspond to the Mahalanobis distance between the object hypothesis x_i and the measurement z_j if the distance is below a certain threshold (e.g., 3). The entry d_{ij} is set to infinity (extremely large value) in case the MHD increases the threshold (forbidden assignment) taking the gating condition into account. The algorithm searches for the lowest sum of matrix entries, so that each row and each column is represented once. Matlab or C implementations for solving the optimization problem can be found, for example, in [64].

The objects' state estimations as well as their measurements underlie spatial and temporal uncertainties. The computation of a guaranteed, correct solution for all situations is impossible. On the one hand, objects may disappear and hence, no further measurements will be received from them. On the other hand, detections can drop, and although the object still exists, the measurement data does not include a detection of the object. Moreover, false detections (clutter) or objects that have not been tracked yet may induce erroneous conclusions. Objects that have been initiated by false detections can be assigned to sporadically appearing false alarms or can take the measurements from correct object representations.

The state uncertainty of objects that have been initiated by a false detection can become very high due to missing associated detections. The statistic distance d^{mh} between the measurement and the false object may become smaller than for the correct object then, although the Euclidean distance is smaller for the latter. This behavior fosters wrong associations, especially when the current prediction of the real object is very distant from the real object state and/or the current realization of the true mea-

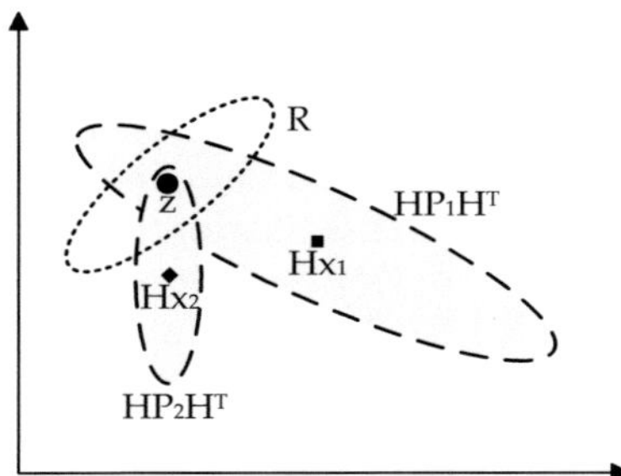

Figure 3.2 Ambiguous association possibilities between objects x_i and measurement z. Although the minimal Euclidean distance is between measurement z and object x_2, the minimal statistical distance lies between measurement z and object x_1.

surement underwent a strong realization. Figure 3.2 shows an example for the described association scenario with two objects and one measurement in the sensor space. The Mahalanobis distance between measurement z and object x_2 is smaller than for object x_1, although the Euclidean distance between measurement z and object x_1 is smaller. Consequently, an association approach is required that considers these effects, which is represented by the JIPDA.

3.2 JIPDA-based Object Tracking and Existence Estimation

3.2.1 Probabilistic Data Association

The previously enumerated effects — missing detections, false detections and spatial vicinity of object and measurement — have to be considered in data association at the same time. A possible solution is provided by the probabilistic multi-object data association with integrated estimation of existence (Joint Integrated Probabilistic Data Association, JIPDA). Probabilistic data association filters utilize probabilistic data association weights β_{ij} instead of a binary ones.

The basic idea of all probabilistic data association approaches is the computation of association probabilities between all N object hypotheses $x_1, \ldots, x_N$ and all M measurements $z_1, \ldots, z_M$ instead of computing binary decision values in the association matrix $\mathbf{A}$ with entries a_{ij}:

$$a_{ij} = \beta_{ij} = p(x_i \leftrightarrow z_j). \tag{3.32}$$

If a missing detection is assumed and only the predicted state estimation is available, the weight β_{i0} for $j = 0$ is computed. Avoiding hard decisions can reduce the impact of false associations and can thereby improve the state estimation. The corresponding association process will be described in Section 3.2.2. The recursive state estimation procedure is similar to the Kalman filter and can be divided into the two phases prediction and innovation. The basic difference is that the soft association decisions evoke the state innovation to be based on the whole set of measurements $\mathcal{Z}_k$ received at time point k instead of being based on a single measurement. The measurements are weighted with the data association weights β_{ij} leading to a new PDF:

$$p(\mathbf{x}_{i,k}|\mathbf{z}_1,\ldots,\mathbf{z}_M) = \sum_{j=0}^{M} \beta_{ij}\tilde{p}_j(\mathbf{x}_{i,k}|\mathbf{z}_j), \tag{3.33}$$

where $\tilde{p}_j(\mathbf{x}_{i,k}|\mathbf{z}_j)$ are the results of the state innovation with the corresponding measurement z_j (compare Equation 3.23 and Equation 3.26). The predicted state estimation for the case of a missing detection is the PDF $\tilde{p}_j(\mathbf{x}_{i,k}|x_0) = p(\mathbf{x}_{i,k|k-1})$.

Unfortunately, the multi-modal distribution (Equation 3.33) is not Gaussian in general. Thus, the posterior PDF needs to be approximated by a single Gaussian distribution to keep the distribution properties. Therefore, innovation hypotheses $\mathbf{x}_{ij}$ are computed for single associations as in Equation 3.23. These single associations serve as auxiliary quantities for the final state update:

$$\mathbf{x}_{ij} = \hat{\mathbf{x}}_{i,k|k-1} + \mathbf{K}_{ij}\gamma_{ij}, \quad i = 1,\ldots,N; \; j = 0,\ldots,M, \tag{3.34}$$

where the Kalman gain and the residuum for missing detections ($j = 0$) are set to $\mathbf{K}_{i0} = \mathbf{0}$ and $\gamma_{i0} = \mathbf{0}$. The preliminary state estimations $\mathbf{x}_{ij}$ will be weighted and accumulated later. Moreover, they enable the computation of the probabilistic association weights β_{ij}. A prerequisite for their computation is that each measurement results from only one object and each object causes only one measurement. Furthermore, the state-dependent detection probability $p_{\mathrm{D}}(\mathbf{x}_i)$ has to be known. The concrete computation of the association weights β_{ij} is described in Subsection 3.2.4.

The update of the object state estimation and its state covariance matrix utilizes the weighted mean of all innovation hypotheses:

$$\hat{\mathbf{x}}_{i,k|k} \;=\; \sum_{j=0}^{M} \beta_{ij}\mathbf{x}_{ij}, \tag{3.35}$$

$$\hat{\mathbf{P}}_{i,k|k} \;=\; \sum_{j=0}^{M} \beta_{ij}[\hat{\mathbf{P}}_{i,k|k-1} - \mathbf{K}_{ij}\mathbf{S}_{ij}\mathbf{K}_{ij}^{\mathrm{T}} + \epsilon_{ij}], \tag{3.36}$$

where ϵ_{ij} is the hypothesis error correcting the approximation error due to unimodal modeling of the multi-modal distribution (in eq. 3.33) and the uncertainty that results from the association alternatives. It is computed by:

$$\epsilon_{ij} = (\mathbf{x}_{ij} - \hat{\mathbf{x}}_{i,k|k})(\mathbf{x}_{ij} - \hat{\mathbf{x}}_{i,k|k})^{\mathrm{T}}. \tag{3.37}$$

3.2.2 Existence Estimation

The computation of the probabilistic association weights β_{ij} in Equation 3.33 is based on the objects' PoEs for the JIPDA filter. Similarly to the state estimation, a Markov chain consisting of a prediction and a innovation step models the PoE. The existence of an object is defined as follows:

Definition 3.1 *An object is denoted as **existent** in the environment model if it is **really present** and it is **relevant** for the vehicle environment perception due to its object class and it is located within a defined tracking region around the ego vehicle.*

The probability of the event that object x exists at time point k is given by

$$p_k(\exists x) := p(\exists x | X_{k-1}, Z_{1:k}) = p(\exists x | x_{k-1}, Z_{1:k}), \tag{3.38}$$

where the Markov property has been applied, as the PoE of an object is expected to depend only on the previous time step. The PoE provides a value $p(\exists x_i)$ between 0 and 1. This value can be interpreted in such way that objects with the same history of attributes, measurement vectors and spatial constellations have been real, positive objects in $p(\exists x_i) \cdot 100\ \%$ of all cases. The measure can be validated statistically and is retrieved from a statistical basis for a correctly determined PoE. In the following, all objects get assigned a PoE that can be applied as a quality measure of the object.

Standard Kalman filter approaches and other filters assume the real existence of the object whose state should be estimated. They apply a measure of quality for each object hypothesis x_i for validation. Often the measure is tested against sensor-dependent or application-dependent heuristic threshold values. Some methods for retrieving this quality measure are summarized in [60], for instance the number of successful measurement associations since object initialization, a measure for the covariance of the state estimation or a combination of different measures. First, the obtained values depend on the sensor or the application, second, the heuristic values are often not reliable.

Additionally, a probabilistic quality measure for the existence of an object, such as a reliable PoE, can provide a great benefit for later system applications where only one object interface would be required. For example, safety-relevant functions that actively control the vehicle require higher PoEs than functions that only warn or inform the driver. Furthermore, application-independent sensor models without thresholds would enable the implementation of a generic sensor fusion framework.

All objects with a PoE higher than any definable threshold ζ_{del} can be tracked until their PoE falls below this threshold. The choice of the threshold ζ_{del} influences the detection ability of the total system. If ζ_{del} is chosen small, the probability of detection is high (less falsely suppressed object hypotheses). On the other hand, this leads to an increase in computational cost, as more hypotheses need to be tracked at the same time.

The JIPDA algorithm for existence estimation consists of the two phases prediction and innovation, just like the Kalman filter for state estimation. The existence of an object is predicted after its prior state and its prior state covariance have been estimated according to Equation 3.16 and Equation 3.17. The existence prediction is based on the state-dependent persistence probability $p_{\mathrm{S}}(x_{k|k-1})$ representing the probability that the object will not disappear, but survive [65]:

$$p_{k|k-1}(\exists x) = p_{\mathrm{S}}(x_{k|k-1})p_{k-1|k-1}(\exists x). \tag{3.39}$$

The persistence probability $p_{\mathrm{S}}(x_{k|k-1})$ enables modeling effects like object disappearance from the defined tracking region around the ego vehicle using the probability $p_{\mathrm{FOV}}(x_{k|k-1})$ and exceeded state constraints with $p_{\mathrm{SC}}(x_{k|k-1})$, e.g., if the object velocity of a pedestrian exceeds 10 m/s, its PoE should be low. Munz [65] additionally modeled mutual occlusion of objects and lowered the PoE of the occluded objects since these are considered as non-existing objects in his definition.

Pedestrians should be tracked instead of cars in this work and occlusion in street crossing scenarios does not mean that these objects do not exist. Instead, the occlusion could be taken into account in the detectability of occluded objects.

A gating method can decrease the computational complexity by limiting the number of potential associations. However, gating can affect the number of available detections. Therefore, it has to be taken into account when computing the association weights β_{ij} and the posterior PoE $p_{k|k}(\exists x)$ based on the complete association hypotheses.

3.2.3 Gating Method

Gating is a method that enables exclusion of unlikely combinations of measurements and objects. It exploits the spatial proximity of object and measurement. Gating regions are defined around the measurement predictions $\hat{z}_{i,k|k-1}$ of each object x_i where the size of the region is based on some distance measure, e.g., the Mahalanobis distance d^{mh} between the predicted measurement and a received measurement. Measurements may only be associated with an object if they are located within the gating region of the object's predicted measurement. The gate size is defined in such way that a measurement of an object is within the gating region with gating probability p_{g}. As in [65], elliptic gating regions with the Mahalanobis distance as distance measure are applied here.

If the measurement residua $\gamma_{ij} = z_j - \hat{z}_{i,k|k-1}$ follow a Gaussian distribution, the squared Mahalanobis distance follows a χ^2-distribution. The gating parameter d^{gate} corresponds to the value up to which one has to integrate this χ^2-distribution with N_{free} degrees of freedom to obtain the gating probability p_{g}. The degree of freedom N_{free} corresponds to the dimension of the measurement space. Increasing degrees of freedom lead to higher values of the gating parameter d^{gate} if the gating probability p_{g} is held constant. The binary gating matrix $\mathbf{G}$ with entries g_{ij} is computed to tell if a measurement z_j is within the gating region of object x_i:

$$g_{ij} = \begin{cases} 1 & , \ (d^{\mathrm{mh}}(x_i, z_j))^2 \leq d^{\mathrm{gate}} \\ 0 & , \ \text{else.} \end{cases} \tag{3.40}$$

Objects are grouped into clusters $\mathcal{C}_{\mathrm{cl}}$, where each cluster contains at least one object (compare [66]). A cluster $\mathcal{C}_{\mathrm{cl}}$ is non-empty if there is beside object x_i at least one additional object x_l competing for the same measurement z_j ($g_{ij} = 1 \wedge g_{lj} = 1$).

The following statements consider only one object cluster $\mathcal{C}_{cl}$ with N^c objects and M^c measurements in it. The number M_j^g represents the number of measurements within the gating region of object x_i.

3.2.4 Computation of Association Hypotheses with Gating

Small gate sizes increase the probability of missing detections (detections that are located outside the gating region). This fact has to be taken into account when computing the probability of the association hypotheses. Furthermore, one has to consider the probability of false detections, that is not spatially or temporally constant in the vehicle environment. One assumes the same prior probability for a false detection for each measurement, so that the expectation value for the number of false detections $\hat{M}^{fa}$ can be computed as in [65]:

$$\hat{M}^{fa} = \sum_{j=1}^{M^c} \prod_{i=1}^{N^c} \left(1 - \frac{p_D(x_i)\, p_g\, p_{k|k-1}(\exists x_i)}{M_j^g} \right)^{g_{ij}}. \tag{3.41}$$

The prior probability for each association hypothesis $p(\mathcal{E}_d)$ is computed based on the expectation value for the number of false detections $\hat{M}^{fa}$. The sets $\mathcal{X}_{na}(e_d)$ and $\mathcal{X}_a(e_d)$ represent sets of objects that have not been detected in the hypothesis $\mathcal{E}_d$ and that have been assigned to a measurement, respectively:

$$\begin{aligned}
p(\mathcal{E}_d) \;=\; &\eta^{-1} \cdot \prod_{x_i \in \mathcal{X}_{na}(e_d)} \left(1 - p_D(x_i) p_g p_{k|k-1}(\exists x_i) \right) \\
&\cdot \prod_{x_i \in \mathcal{X}_a(e_d)} \left(\frac{V_d}{\hat{M}^{fa}} \Lambda(z_{m(i,d)}|x_i) p_D(x_i) p_g p_{k|k-1}(\exists x_i) \right),
\end{aligned} \tag{3.42}$$

where $z_{m(i,d)}$ is the measurement that is assigned to object x_i within the association hypothesis $\mathcal{E}_d$ and η is a normalization constant. It is computed in such way that the probabilities of all association hypotheses $p(\mathcal{E}_d)$ sum up to 1. V_d denotes the volume of the gating region of all objects in the cluster and $\Lambda(z_j|x_i)$ represents the prior probability density for measurement z_j [58,60]:

$$\Lambda(z_j|x_i) = \frac{1}{p_g} \int \mathcal{N}(\mathbf{z}_j, \mathbf{H}_j\hat{\mathbf{x}}_i, \mathbf{R}_j)\mathcal{N}(\mathbf{x}_i, \hat{\mathbf{x}}_i, \hat{\mathbf{P}}_i)\, \mathrm{d}\mathbf{x}_i. \tag{3.43}$$

Both the case of a missing detection for an existing object and the case that a measurement results from a non-existing object have to be considered. The introduction of pseudo-measurements as in [60, 65] simplifies the notation. The set of measurements is enlarged by two symbols to represent missing detections $\emptyset$ and non-existence of an object $\sharp$, so that $\mathcal{Z}^* = \{z_1,\ldots,z_M,\emptyset,\sharp\}$. Analogously, a clutter source that is represented by the symbol $\copyright$ and the symbol $\mathbf{b}$ for a new born object are added to the set of objects ($\mathcal{X}^* = \{x_1,\ldots,x_N,\copyright,\mathbf{b}\}$). A pair $e = (x \in \mathcal{X}^*, z \in \mathcal{Z}^*)$ represents a single association, while the set $\mathcal{E} = \{e_i\}$ constitutes a complete association. The special elements $\emptyset, \sharp, \copyright$ and $\mathbf{b}$ may be multiply used, whereas the other elements from the sets $\mathcal{X}^*$ and $\mathcal{Z}^*$ may only be used once.

Each association hypothesis $\mathcal{E}$ consists of elementary assignments e between elements representing measurements or missing detections and objects, clutter sources, or newly born objects. The single probability that an object x_i exists but could not be observed is:

$$p(e = (x_i, \emptyset)) = \frac{(1 - p_D(x_i)\, p_g)\, p_{k|k-1}(\exists x_i)}{1 - p_D(x_i)\, p_g\, p_{k|k-1}(\exists x_i)} \sum_{\mathcal{E}:(x_i,\emptyset)\in\mathcal{E}} p(\mathcal{E}), \qquad (3.44)$$

whereas the probability that an object x_i exists and a measurement z_j results from that object can be written as:

$$p(e = (x_i, z_j)) = \sum_{\mathcal{E}:(x_i,z_j)\in\mathcal{E}} p(\mathcal{E}). \qquad (3.45)$$

Thus, the posterior PoE for object x_i can be computed by summarizing the probabilities of all elementary assignments that assume the existence of object x_i:

$$p_{k|k}(\exists x_i) = p(e = (x_i, \emptyset)) + \sum_{j: g_{ij}=1} p(e = (x_i, z_j)). \qquad (3.46)$$

Consequently, the data association weights for each combination of object and measurement can be determined as the ratio of the corresponding elementary, single assignment and the posterior PoE of the corresponding object x_i:

$$\beta_{ij} = \frac{p(e = (x_i, z_j))}{p_{k|k}(\exists x_i)} \text{ for } j > 0, \qquad (3.47)$$

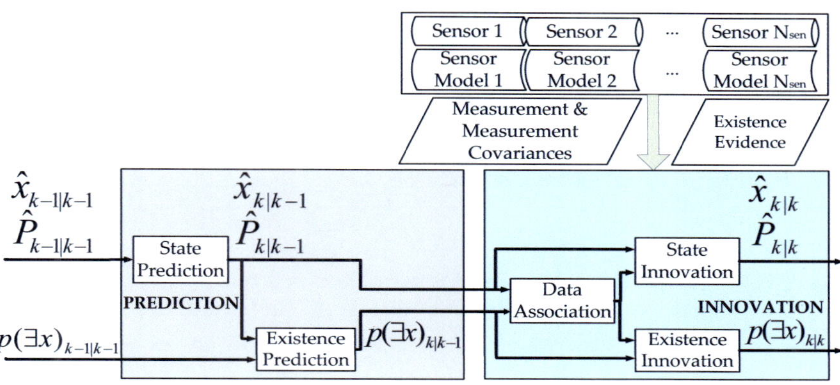

Figure 3.3 Principle of state and existence estimation in JIPDA filtering.

$$\beta_{ij} = \frac{p(e = (x_i, \varnothing))}{p_{k|k}(\exists x_i)} \text{ for } j = 0. \tag{3.48}$$

Finally, Equation 3.35 and Equation 3.36 update the object's state estimation and its covariance. Figure 3.3 summarizes the principle of state and existence estimation of the resulting EKF with JIPDA (EKF-JIPDA).

3.2.5 Hypotheses Tree for the JIPDA Implementation

An efficient algorithm is required for the computation and the enumeration of the association hypotheses. A realizable, intuitive and extensible implementation is provided by a graph-based realization using a hypotheses tree as in [60, 65, 67]. The hypotheses tree enables a systematic enumeration of all association hypotheses.

The path $e_0, \ldots, e_{L(d)}$ from the root node e_0 to leave node $e_{L(d)}$ represents an association hypothesis $\mathcal{E}_d = \{e_0, \ldots, e_{L(d)}\}$ that is independent of the order, see Figure 3.4 for the illustration of a simple hypothesis tree with two objects and two measurements. The product of the probabilities of the single assignments returns the probability of the association hypothesis:

$$p(\mathcal{E}_d) = \prod_{e \in \mathcal{E}_d} p(e). \tag{3.49}$$

If a hypothesis $\mathcal{E}_d$ contains the single assignment $e = (x_i, z_j)$ where measurement z_j is associated with object x_i, the hypothesis $\mathcal{E}_d$ belongs to the

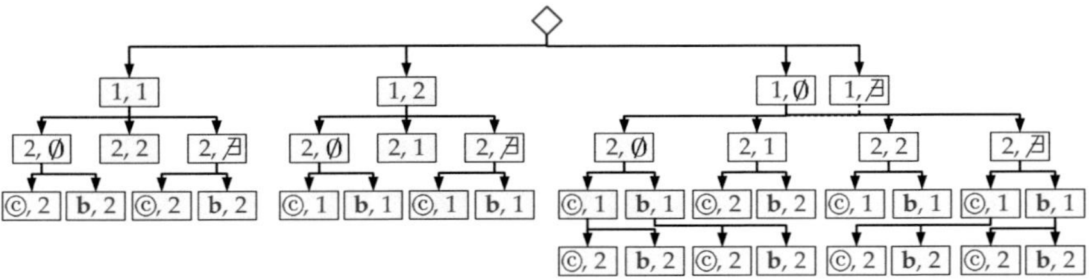

Figure 3.4 JIPDA Hypotheses Tree for two objects and two measurements: Each node contains one element from set $\mathcal{X}^*$ (in front of comma) and one from $\mathcal{Z}^*$.

set of true positive associations $\mathcal{E}_{ij}^{\mathrm{TP}}$ ($\mathcal{E}_d \in \mathcal{E}_{ij}^{\mathrm{TP}}$). Furthermore, if the hypothesis $\mathcal{E}_d$ expects that x_i exists (although it might not have been detected) and does therefore not contain the single assignment $e = (x_i, \not\exists)$, the hypothesis $\mathcal{E}_d$ belongs to the set $\mathcal{E}_i^{\exists}$ for object existence ($\mathcal{E}_d \in \mathcal{E}_i^{\exists}$). Marginalization leads to the posterior PoE of object x_i [67]:

$$p_{k|k}(\exists x_i) = \frac{\sum_{\mathcal{E} \in \mathcal{E}_i^{\exists}} p(\mathcal{E})}{\sum_{\mathcal{E}} p(\mathcal{E})}. \tag{3.50}$$

Analogously, the association weights β_{ij} can be determined by:

$$\beta_{ij} = \frac{\sum_{\mathcal{E} \in \mathcal{E}_{ij}^{\mathrm{TP}}} p(\mathcal{E})}{\sum_{\mathcal{E} \in \mathcal{E}_i^{\exists}} p(\mathcal{E})}. \tag{3.51}$$

Lookup tables can serve for an efficient computation of the hypotheses probabilities. The basic elements are stated by the node probabilities. Mählisch [60] proposes a computation rule dependent on the type of association. There, the inference probability $p_{\mathrm{TP}}(z_j)$ represents the probability that measurement z_j results from a real object. The probability is estimated from the specific measurement values and features. The counter event is the false alarm probability $p_{\mathrm{FP}}(z_j) = 1 - p_{\mathrm{TP}}(z_j)$. Thereby, sensor-specific existence evidence is directly incorporated in the JIPDA filter [65]. The normal distribution $\mathcal{N}(z_j, \hat{z}_j, \mathbf{R}_j)$ represents the spatial uncertainty of measurement z_j. The likelihood probability $p_\Lambda(x_i, z_j)$ models the elementary impact of the spatial proximity between measurement and object. It is based on the Mahalanobis distance and takes into account that an object x_i is more likely to exist if there is a measurement z_j close to it:

$$p_\Lambda(x_i, z_j) = \exp(-\frac{1}{2}\gamma_{ij}^{\mathrm{T}} \mathbf{S}_{ij}^{-1} \gamma_{ij}).$$

(3.52)

The detection probability $p_{\mathrm{D}}(x_i)$ depends on the object state and enables modeling sensory blind regions within the defined tracking region around the ego vehicle and sensor-dependent detection limitations.

The probability of object birth p_{B} models the probability that a measurement results from an unobserved object. There are three kinds of birth processes:

- First detection, since the source of a measurement has just entered the defined tracking region around the vehicle or is located within the detection range for the first time.

- An object has just left the region where it was occluded by another tracked or irrelevant object, and no object has been tracked there so far.

- Detection of objects that have not been detected earlier due to missing detections, dynamic pitching of the ego vehicle or due to other effects.

If this birth probability is high enough, a new object hypothesis is instantiated. Each sensor receiving a measurement z_j has to generate a new object hypothesis $x_j^{\mathcal{H}} \sim \mathcal{N}(\mathbf{x}_j^{\mathcal{H}}, \hat{\mathbf{x}}_j^{\mathcal{H}}, \mathbf{P}_j^{\mathcal{H}})$. However, the new object hypothesis is not considered for the state estimation of already tracked objects.

The probability of object birth is decomposed into three components. A sensory part constitutes the first component and it is represented by the sensory inference probability of the potential object $p_{\mathrm{TP}}(z_j)$ based on the properties of measurement z_j. The second component can take into account map data, so that no objects can be born, e.g., if they are behind a wall. As no map data is available here, this component is not modeled in this work. The third component depends on the current environment model that is compared to object hypotheses resulting from measurements. It considers the spatial uncertainty of objects existing in the environment model and their PoE.

Since an integrated description of the state uncertainty and of the existence uncertainty of each object and each object hypothesis is required, a probabilistic hypothesis density function (PHD) is defined based on the

FISST theory from [52]. The JIPDA method corresponds to a representation of the Multi-Bernoulli distribution of the FISST modeling [60]. Therefore, the Multi-Bernoulli PHD defined on objects x_i can be expressed by

$$\mathcal{D}_k(x_i) = p_{k|k}(\exists x_i)\mathcal{N}(\mathbf{x}_{i,k}, \hat{\mathbf{x}}_{i,k|k}, \hat{\mathbf{P}}_{i,k|k}), \tag{3.53}$$

where a derivation can be found, e.g., in [65]. The integral over the Multi-Bernoulli PHD function of an object x_i corresponds to the object's PoE. The PHD function is only defined for point objects without dimension or orientation — information that is relevant in many automotive applications of environment perception. Munz [65] presented a compact representation for car objects. Here, the approach is slightly different. The horizontal dimension of a pedestrian is defined as square ($\text{square}_{x_i,w}(x,y)$) with edge length w (e.g., 0.5 m for pedestrians) that is shifted to the pedestrian's position. The dimension of a pedestrian could also be represented by a circle with a given radius.

The PDF for the occupation of a position on the x,y-plane by an object is denoted as $p_{\text{occ}}(x,y)$. It is provided by multiplication of the two-dimensional convolution of the distribution of the object position and the shifted square with the PoE of the object [65]:

$$p_{\text{occ}}(x,y) = p_{k|k}(\exists x_i)[\mathcal{N}((x,y)^\mathrm{T}, (\hat{x},\hat{y})^\mathrm{T}, \hat{\mathbf{P}}_i^{xy}) * \text{square}_{x_i,w}(x,y)], \tag{3.54}$$

where $\hat{\mathbf{P}}_i^{xy}$ contains only the first two elements of $\hat{\mathbf{P}}_i$ representing the state covariance of the object position. The time index k is neglected from now on. This PDF can also be interpreted as dimensional probabilistic hypothesis distribution (DPHD) $p_{\text{occ}}(x,y) = \mathcal{D}_{x_i}^{\text{d}}(x,y)$. However, the DPHD $\mathcal{D}_{x_i}^{\text{d}}(x,y)$ does not fulfill an important property of a PHD since it represents a dimensional object instead of a point object that could be modeled by a Dirac distribution. Therefore, the two-dimensional carrier space of the DPHD is separated into small cells. The probability of the occupancy of an area $\mathcal{A}$ of a cell is obtained by integration. The numerical approximation applies lower and upper sums:

$$\tilde{\mathcal{D}}_{x_i}^{\text{d}} = \int_{\mathcal{A}} \mathcal{D}_{x_i}^{\text{d}} \approx \sum_x \sum_y \mathcal{D}_{x_i}^{\text{d}}(x,y) \cdot d_x d_y, \tag{3.55}$$

where d_x and d_y represent the edge lengths of small grid cells in the x,y-plane. The result is denoted as discrete DPHD function $\tilde{\mathcal{D}}_{x_i}^{\text{d}}$ and corresponds to a local occupancy map. The probability of occupancy is obtained directly from the fusion results of the JIPDA method after each fusion cycle.

The discrete, dimensional probabilistic hypothesis density function (DPHD) is dedicated to the simultaneous representation of the object hypotheses and the present objects. One checks if the positions of the potential new object $x_j^{\mathcal{H}}$ overlap with positions of objects from the environment model:

$$\tilde{\mathcal{D}}_{x_j^{\mathcal{H}} \cap \mathcal{X}_{\text{env}}^{\text{d}}}(x,y) = \min\left(1, \sum_{i=1}^{N} \tilde{\mathcal{D}}_{x_j^{\mathcal{H}}}^{\text{d}}(x,y) \cdot \tilde{\mathcal{D}}_{x_i}^{\text{d}}(x,y)\right), \tag{3.56}$$

where $\mathcal{X}_{\text{env}} = \{x_1, \ldots, x_N\}$ represents the set of objects in the environment model. The results of the object overlap per cell are not independent from each other. The distribution is reduced to the mode of $\tilde{\mathcal{D}}_{x_j^{\mathcal{H}} \cap \mathcal{X}_{\text{env}}}^{d}(x,y)$ to enable a simple computation of the overlap probability [65]:

$$p_{\text{I}}(z_j|x_1, \ldots, x_N) = \max(\tilde{\mathcal{D}}_{x_j^{\mathcal{H}} \cap \mathcal{X}_{\text{env}}}^{\text{d}}(x,y)). \tag{3.57}$$

The distribution $p_{\text{I}}(z_j|x_1, \ldots, x_N)$ is an indicator for the probability that there is an overlap between the object hypothesis generated by measurement z_j and at least one present object from the environment model. It incorporates uncertainty in the spatial distribution and in the existence estimation. The probability that an object hypothesis and an object in the environment model represent the same physical object is given by

$$p_{\Lambda}^{\mathcal{H}}(z_j|x_i) = \exp\left(-\frac{1}{2}(\hat{\mathbf{x}}_j^{\mathcal{H}} - \hat{\mathbf{x}}_i)^{\text{T}}(\hat{\mathbf{P}}_j^{\mathcal{H}} + \hat{\mathbf{P}}_i)^{-1}(\hat{\mathbf{x}}_j^{\mathcal{H}} - \hat{\mathbf{x}}_i)\right), \tag{3.58}$$

where $\hat{\mathbf{P}}_i$ is the covariance matrix of x_i and $\hat{\mathbf{P}}_j^{\mathcal{H}}$ is the covariance matrix of the new object hypothesis $\mathbf{x}_j^{\mathcal{H}}$. The equation evaluates the Mahalanobis distance between the object x_i from the environment model and the new object hypothesis.

Finally, the probability of birth based on the current environment model and a received measurement z_j is given by

$$p_{\text{B}}(z_j|x_1, \ldots, x_N) = (1 - p_{\text{I}}(z_j|x_1, \ldots, x_N)) \prod_{i=1}^{N}(1 - p_{\Lambda}^{\mathcal{H}}(z_j|x_i)). \tag{3.59}$$

It is possible to compute the probabilities of the single assignments (node probabilities) based on the described probabilities where one formula is provided for each node type.

True positive association node (TP):

$$p(e = \{x_i, z_j\}) = p_{k|k-1}(\exists x_i) \cdot p_{\text{TP}}(z_j) \cdot p_{\text{D}}(x_i) \cdot p_{\Lambda}(z_j|x_i) \cdot p_{\text{g}}. \qquad (3.60)$$

False positive association node (FP):

$$p(e = \{\copyright, z_j\}) = (1 - p_{\text{TP}}(z_j)). \qquad (3.61)$$

False negative association node (FN):

$$p(e = \{x_i, \varnothing\}) = p_{k|k-1}(\exists x_i) \cdot ((1 - p_{\text{D}}(x_i)) + p_{\text{D}}(x_i) \cdot (1 - p_{\text{g}})). \qquad (3.62)$$

True negative association node (TN):

$$p(e = \{x_i, \not\exists\}) = 1 - p_{k|k-1}(\exists x_i). \qquad (3.63)$$

Birth association node:

$$p(e = \{\mathbf{b}, z_j\}) = p_{\text{TP}}(z_j) \cdot p_{\text{B}}(z_j|x_1, \ldots, x_N). \qquad (3.64)$$

The computation of the probabilities for the complete association hypotheses of measurements to objects builds the basis of the association method. The JIPDA method gets more and more complex with an increasing number of objects or measurements due to the manifold combinatorics. The computational effort can be reduced enormously if gating is applied to exclude very unlikely combinations prior to the computation of the associations.

3.2.6 Object Management in JIPDA Filtering

The object management instantiates new object hypotheses, initializes object states, and manages the object list efficiently. If the birth probability $p_{\text{B}}(z_j)$ exceeds the threshold value ζ_B, the object hypothesis $x_j^{\mathcal{H}}$ resulting from the measurement z_j of the sensor is added to the object list which is referred to as instantiation. Otherwise, the object hypothesis is rejected.

Initialization of an object state means the identification of initial state values. Sensors often allow a state initialization only with high uncertainties leading to big gating regions after the prediction step. Thus, the effort for data association becomes high due to complex ambiguities and the computational demand increases. Multi-step approaches are advantageous, especially for dynamic state variables. Static state variables that cannot be observed by the sensor (e.g., the components of the object dimension) are initialized using prior values from statistics. A new object hypothesis is included within the environment model with the status *uninitialized* in case of the multi-step approach. An initialization is accomplished in one of the next measurement cycles after data association with the new sensor data.

The initialization considers only measurements that result very likely from unobserved objects. The required association weight β_{bj} is computed within the JIPDA based on the birth model. In this case, a nearest neighbor association approach usually suffices for the resolution of ambiguities since only a few candidates exist for initialization. Object speeds and object sizes have to be within plausible ranges. Therefore, the object has to pass these consistency tests before final object initialization. If all conditions are met, the initial PoE $p(\exists x_j)_{\text{init}}$ of the object hypothesis is computed according to

$$p(\exists x_j)_{\text{init}} = p_{\text{TP}}(z_j) p_{\text{B}}(z_j | x_1, \ldots, x_N) p_{\text{fasso}}(i, j), \qquad (3.65)$$

where $p_{\text{fasso}}(i, j)$ is the probability of a false association between measurement z_j and another object x_i. It can be chosen as constant prior probability that has been determined statistically.

An alternative approach for object initialization, called multi-hypothesis initialization (MHI) is described in [65]. It initializes multiple object hypotheses $x_{i1}, \ldots, x_{iK}$ at once. Thereby, the PoEs of objects in a group are explicitly coupled and a normalization of the PoEs of all objects belonging to the group has to be accomplished after prediction and innovation using normalization weights. The choice of normalization weights may cause an amplified effect of the PoEs versus 1 or 0 for object hypotheses that contradict each other. Therefore, the first approach is applied here.

3.2.7 Algorithm Complexity and Real-Time Computation

The runtime of the presented JIPDA method is within the complexity class $O(N^M)$ if a uniform measure of complexity is chosen. The runtime increases exponentially with the number of measurements M (within a cluster). It is assumed for simplicity reasons that $N \leq M$. This means that at least as many measurements are received as objects are present in the current environment model. The assumption is reasonable for most sensors. If $M < N$, the complexity would be $O(M^N)$ and all considerations could be accomplished analogously.

The number of required computation steps is proportional to the number of nodes in the hypotheses tree. The runtime of the approach is basically influenced by the computation of hypotheses in the tree, and thus, it increases exponentially with the number of measurements and currently tracked objects, respectively. An upper bound for the computation of an iteration has to be guaranteed for the implementation in an electronic control unit independently of the complexity of the currently observed scenario. Therefore, the method requires an adaption. There are some JPDA (Joint Probabilistic Data Association) approximation methods in the literature that compute the data association probabilities β_{ij} directly without enumerating the events. Several authors proposed approaches for a reduction of the computational effort. A short summary can be found in Chapter 4.1.2. Here, the adaptive procedure for a reduction of the hypotheses number as proposed by Munz [65] is applied. The reduction of contradictory hypotheses enables a guaranteed maximum runtime where only alternative hypotheses on the state level are excluded. All alternatives (non-existence, missing detection, false detection, object birth and correct detection) are considered for the estimation of existence. Therefore, the applied gating method is adapted in a way that gating regions for objects are reduced in measurement regions where many gating regions overlap. These regions would cause complex hypotheses trees due to the high combinatoric diversity. This adaption is accomplished until the complexity of the problem is reduced sufficiently to enable algorithm termination within the given maximum time. The gating probability p_g enables an direct adaption of the gating regions. The JIPDA method degenerates to an IPDA (Integrated Probabilistic Data Association) method with local nearest neighbor association in the extreme case. A missing measurement due to the reduced gating region is then interpreted as a missing detection.

3.3 Classification and Parameter Estimation

Classification techniques are applied to distinguish between different maneuver types of the ego vehicle in the situation evaluation section. Classification is a technique for the prediction of group membership for data instances on the basis of a training set of data containing observations whose group membership is known. The target variable of the learning problem may only take on a small number of discrete values. Otherwise, if the target variable that should be predicted is continuous, it is called parameter estimation. While parameter approaches like linear regression use a finite number of parameters, the number of parameters grows linearly with the number of training samples N_t in non-parametric approaches, such as locally weighted regression.

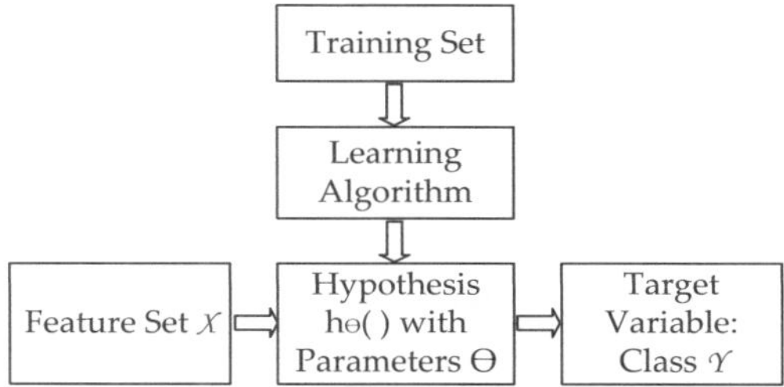

Figure 3.5 Principle of supervised learning: A training data set serves the optimization of the parameters Θ of the hypothesis function h_Θ. Given the features $\mathbf{x} \in \mathcal{X}$ as input variables, the hypothesis function h_Θ later determines the class of an input sample as the target variable $\mathcal{Y}$.

Machine learning distinguishes between supervised and unsupervised learning.[1] This work focuses on supervised learning.

The principle of supervised learning is depicted in Figure 3.5. A training set $\left\langle \mathbf{x}^{(i)}, y^{(i)} \right\rangle$ with $i = 1, \ldots, N_t$ serves the optimization of the parameters Θ of the hypothesis function $h_\Theta(\cdot)$. Given the input variables $\mathbf{x}$ containing the features, the hypothesis function h_Θ later determines the class of an input sample as the target variable y. More formally, a given training set should enable to learn a function $h_\Theta : \mathcal{X} \mapsto \mathcal{Y}$, so that $h_\Theta(\mathbf{x})$ is a good predictor for the corresponding value of $y \in \mathcal{Y}$. An important expression in context with classification and parameter estimation is the likelihood $\Lambda_\Theta = p(y|\mathbf{x}, \Theta)$. Its maximization with respect to the parameters Θ serves finding the parameters of the hypothesis function $h_\Theta(\cdot)$ for a given training data set. The parameters Θ should be chosen in such

---

[1]Unsupervised learning algorithms are usually applied when it is difficult to provide explicit supervision (unambiguous labels for every input) to a learning algorithm in sequential decision making and control problems. Then, one only provides a reward function to the algorithm that indicates to the learning agent when it is doing well or badly (reinforcement learning). It is the learning algorithm's responsibility to find the actions over time that give the largest rewards. Reinforcement learning has been successfully applied in diverse fields such as autonomous helicopter flight, cell-phone network routing, marketing strategy selection or factory control.

way that the observed results become most probable based on the given model.

Moreover, an important distinction between learning algorithms is made by the kind of model they rely on, such as linear and non-linear models. For example, Generalized Linear Models (GLM) are based on distributions from the exponential family, such as the Gaussian, the Bernoulli or the Poisson distribution. Furthermore, one distinguishes between discriminative and generative supervised learning algorithms. While the first try to learn $p(y|\mathbf{x})$ directly or try to learn mappings directly from the space of inputs $\mathcal{X}$ to the labels $\{0,1\}$, the latter try to model $p(\mathbf{x}|y)$ and $p(y)$.

Discriminative learning algorithms (such as logistic regression) use a given training set to find a decision boundary between the different classes. The algorithm checks in the classification step of a new sample on which side of the boundary it falls, and makes its prediction accordingly.

Generative learning algorithms try to find a description of known classes using features and the prior probability of a sample belonging to a class (class prior $p(y)$). Then the features of a new sample are checked for similarity against the features of the known classes for classification. After modeling the class priors $p(y)$ and $p(\mathbf{x}|y)$, the algorithm uses the Bayesian rule to derive the posterior distribution on y given $\mathbf{x}$:

$$p(y|\mathbf{x}) = \frac{p(\mathbf{x}|y)p(y)}{p(\mathbf{x})}. \tag{3.66}$$

Common methods for supervised learning are, e.g., support vector machines (SVM) and naive Bayesian classifiers. Linear SVM algorithms are utilized in this work to identify pedestrians in camera images where histograms of oriented gradients (HOG) build the features. A naive Bayesian classifier serves the classification of driving maneuvers of the ego vehicle — such as lane following, lane change or turn, see Chapter 5.2.

SVMs belong to the class of non-linear classifiers with a discriminative model and have usually a higher predictive accuracy than naive Bayesian classifiers that are based on a generative model. However, the latter are normally easier to interpret and use Equation 3.66. The prediction speed of Bayesian classifiers is very high and their memory usage is low for simple distributions, although the speed may be lowered and the memory usage may be increased by complex kernel distributions. SVMs show good properties for prediction speed and memory usage when there are only a few support vectors.

Dynamic Time Warping (DTW) or Hidden Markov Models (HMM) are other exemplary approaches for behavior understanding that are not used here. DTW is a template-based dynamic programming matching technique. HMMs outperform DTW in the processing of undivided successive data. Bayesian networks incorporate the advantage to use prior knowledge and to model dynamic dependencies between state parameters, also these dependencies make the model design and its computation much more complex.

Support Vector Machines

SVMs are based on the idea of large margins and are one of the most prevalent applications of convex optimization methods in machine learning. One defines separating hyperplanes (decision boundaries) between the classes in such way that the distances of the samples of each class to the hyperplane are maximized. The idea is that the farther a sample is located from the hyperplane on one side, the more confident one is that the sample belongs to the class on the corresponding side of the hyperplane. Thus, one tries to maximize the distance of the samples to the hyperplane — the so-called margin. A functional margin is related to the geometric margin via normalization. The class labels for binary classification problems are usually denoted by $y \in \{-1, 1\}$ instead of $\{0, 1\}$ in context with SVMs. The feature points with the smallest margins to the decision boundary are the so-called support vectors. The number of support vectors can be much smaller than the size of the training set. SVMs can learn high dimensional feature spaces when one chooses a kernel formulation where the kernel can be computed efficiently. Since the development of the pedestrian detection and classification algorithm is not part of this work and only the results are used and evaluated, the SVM algorithms are not presented in more detail here. An introduction to SVMs can be found, e.g., in [68].

Naive Bayesian Classifier

The classification of e-mails into spam and non-spam is a popular application of the naive Bayesian classifier. The classifier is MAP-based and therefore uses a generative model where one aims to maximize the right hand side of Equation 3.66:

$$\hat{y} = \arg\max_{y} p(y|\mathbf{x}) = \arg\max_{y} \frac{p(\mathbf{x}|y)p(y)}{p(\mathbf{x})} = \arg\max_{y} p(\mathbf{x}|y)p(y). \quad (3.67)$$

If one models $p(\mathbf{x}|y)$ and $p(y)$ to make a prediction, it is not necessary to calculate the denominator since it is independent of y. However, the parameter vector will end up with too many parameters in problems with high-dimensional feature vectors $\mathbf{x}$ if the feature vector is modeled explicitly with a multinomial distribution over all possible outcomes (feature combinations). Moreover, this requires giant data sets for training. Therefore, one applies the naive Bayesian assumption that expects conditional independence of all features $\mathbf{x}(i)$ given y, e.g., $p(\mathbf{x}(k_1)|y) = p(\mathbf{x}(k_1)|y, \mathbf{x}(k_2))$. Even though the Bayesian classifier uses an extremely strong assumption, it works well on many real world problems — even when the assumptions are not met. Laplace smoothing is usually applied if the training set does not contain all possible features with all states, so that $p(\mathbf{x}(i)|y(j) = z) = 0$ and thus, the class posterior probabilities result in $0/0$. The algorithm cannot provide a prediction anymore. The assumption that the probability of an event that has not been seen in the finite training data set is 0 might be statistically misleading. Therefore, Laplace smoothing avoids ending up with zeros by setting $p(\mathbf{x}(i)|y(j) = z)$ to a very small value and subsequent normalization of all probabilities. The naive Bayesian classifier often works well and it is a good starting point due to its simplicity and ease of implementation.

3.3.1 Model Selection and Model Evaluation

When designing a classification algorithm, one has to think of an appropriate model for the given problem. The model choice should aim for a minimization of the *generalization error of a hypothesis*. The generalization error of a hypothesis is the expected error on examples that were not necessarily included in the training set. On the one hand, a model A might fail to accurately capture the data structure and under-fits the data even though there is an infinitely large amount of training data which results in an error called *bias*. On the other hand, another model B might perfectly fit to the examples within the training set but does not necessarily provide good predictions for additional samples. There is a large risk that the model does not reflect the wider pattern of the relationship between $\mathbf{x}$ and y, although it fits patterns in the small, finite training set very well. This leads to the second component of the generalization error consisting of the *variance of the model fitting procedure*. There is a trade-off between bias and variance, so that a third model C — something in the middle — might perform better than the extremes A and B.

Empirical risk minimization aims for minimization of the generalization error. Therefore, one defines the training error for a hypothesis h_Θ as the fraction of misclassified training examples. The generalization error is then defined as the probability that h_Θ will misclassify a new example $(\mathbf{x}, y)$ from the same distribution. Empirical risk minimization picks the hypothesis function h_Θ that leads to the smallest training error from the class of hypothesis functions $\mathcal{H}$. The training error is close to the generalization error with high probability if the size of the training data set N_t is large. If a larger class of hypothesis functions is used, the bias will decrease but the variance will increase, see [69] for evidence. If the hypothesis class is infinite and has N_p parameters, the number of training examples required for an algorithm trying to minimize the training error is usually roughly linear in the number of parameters N_p of class $\mathcal{H}$.

If the classes in the data are not separable, there will always be a classification error. However, one aims to keep the consequences of misclassification as low as possible. The required evaluation criteria to numerically express the consequences of a classifier decision can be a cost matrix $\mathbf{C}$ where the elements c_{ij} represent the costs that are created when a sample belongs to class j but the classifier decides for class i. The classifier performance can be evaluated by analysis of the mean costs where the classification results from a validation training set are related to the elements of the cost matrix $\mathbf{C}$

$$\hat{\epsilon}(h_\Theta) = \sum_{i=1}^{N_{\text{class}}} \sum_{j=1}^{N_{\text{class}}} c_{ij} \cdot e_{ij}, \tag{3.68}$$

where an element e_{ij} of matrix $\mathbf{E}$ represents the fraction of samples that were assigned to class i and belonged to class j. Thus, the matrix diagonal ($i = j$) contains the fractions of correct classification. The elements of the cost matrix $\mathbf{C}$ may depend upon the problem and require further knowledge about the model.

Hold-out cross-validation (simple cross-validation) and N_s-fold cross-validation are popular methods to select the model with the best performance from a finite set of models $\mathcal{M} = \{M_1, \ldots, M_{N_p}\}$. The cross-validation algorithms are applicable for model selection as well as for evaluation of a single model or algorithm. In this work, 10-fold cross-validation is applied for training and evaluation of the maneuver classifier. The principle is described in Appendix A.3 or in [69].

Feature selection is a special case of model selection. It is useful to find meaningful features and the right number of features for the learning task. If there are N_f features, then the feature selection can be considered as a model selection problem with 2^{N_f} possible models, since every feature can be relevant or less important. Wrapper model feature selection aims to find the feature subset with the smallest generalization and can be applied with forward search or with backward search. The first starts with one feature and enlarges the feature subset until more or less all possible combinations have been tested, while the latter starts with a set of all features and reduces this set until there are no more features in it. Both algorithms are quite expensive and complete forward search takes about $O(N_f^2)$ calls to the learning algorithm.

3.4 Discrete Event Systems and Hybrid Automatons

Discrete (Dynamic) Event Systems (DES) model dynamics that cannot be captured by differential equations or by difference equations. The word *discrete* does not mean that the time or the state have to be discrete. It rather refers to the fact that the dynamics are made up of events, although these events can evolute continuously. In this work, DESs are used to model and to predict the behavior of the ego vehicle and the behavior of detected road users in its environment.

DES that combine time-driven dynamics with event-driven dynamics are referred to as hybrid systems. One can associate a set of differential equations describing the evolution of the continuous variables of interest to each discrete state of the system. A hybrid system is capable to describe switching dynamics and exhibits both continuous and discrete dynamic behavior. The most common modeling framework for hybrid systems is provided by the hybrid automaton. It is imaginable as an extension of a timed automate with guards where arbitrary time-driven dynamics at each discrete state characterize one or more continuous state variables.

The continuous state $\mathbf{x}$ may take values within continuous sets whereupon only a single initial discrete state q from the set of discrete states ($q \in \mathcal{Q}$) is assumed. In the case of several initial discrete states (modes), the analysis of reachable states can be performed for each mode separately. Thus, the system state can be expressed by $(q, \mathbf{x})$ with $\mathbf{x} \in \mathcal{X}$ (usually $\mathcal{X} \subseteq \mathbb{R}^n$).

The state evolution of the hybrid automaton starts in the initial mode q^0 and in an initial state $\mathbf{x}(0) \in \mathcal{X}^0$. The mode specific flow function f_{flow} describes the development of the continuous state. As long as the continuous state is within a guard set $\mathcal{G}$, the corresponding transition may be taken. It has to be taken if the state would leave the invariant $\text{inv}(q_i)$. When the transition from the previous mode q_i to the next mode q_j is taken, the system state is updated according to the jump function h_{jump} and the flow function f_{flow} within the next invariant $\text{inv}(q_j)$. The transitions between discrete states are forced by exogenous events from the outside and endogenous events that occur when a time-driven state variable enters a particular set and violates an invariant condition. If the endogenous event evokes a guard condition, an exogenous event is subsequently required to cause a transition related to this guard condition. Figure 3.6 illustrates the principle of the reachable sets in a hybrid automaton. Here, the hybrid automaton is defined based on the definition in [70, 71] with additional consideration of uncertain parameters and restrictions on jumps and guard sets:

Definition 3.2 *(Hybrid Automaton):*
A hybrid automaton, denoted by G_{ha}, is an eleven-tuple

$$G_{\text{ha}} = (Q, q^0, \mathcal{X}, \text{inv}, \mathcal{X}^0, \mathcal{U}, \mathcal{P}, \mathcal{T}, f_{\text{flow}}, g_{\text{guard}}, h_{\text{jump}}) \qquad (3.69)$$

where
$Q = \{q_0, \dots, q_{N_{\text{mode}}}\}$ is a finite set of discrete states or modes,
q^0 is an initial discrete state,
$\mathcal{X}$ is a continuous state space (usually $\mathbb{R}^n$),
$\text{inv} : Q \mapsto 2^{\mathcal{X}}$ is a mapping resulting in a set that defines an invariant condition (also called domain) for each mode q ($\text{inv}(q) \subseteq \mathcal{X}$),
$\mathcal{X}^0$ is an initial continuous state, so that $\mathcal{X}^0 \subseteq \text{inv}(q^0)$
$\mathcal{U}$ is a set of admissible control inputs (usually $\mathcal{U} \subseteq \mathbb{R}^m$),
$\mathcal{P} \subseteq \mathcal{I}^p$ is the parameter space,
$\mathcal{T}$ is a set of discrete state transitions with $\mathcal{T} \subseteq Q \times Q$ where a transition from $q_i \in Q$ to $q_j \in Q$ is denoted by (q_i, q_j),
$f_{\text{flow}} : Q \times \mathcal{X} \times \mathcal{U} \times \mathcal{P} \to \mathbb{R}^n$ is a flow function defined as vector field for the time derivative of $x : \dot{x} = f_{\text{flow}}(q, x, u, \rho)$,
$g_{\text{guard}} : \mathcal{T} \mapsto 2^{\mathcal{X}}$ is a mapping function that results in a set $\mathcal{G}$ defining a guard condition $\mathcal{G} \subseteq Q \times Q \times \mathcal{X}$ for each transition from q_i to q_j where $g_{\text{guard}}((q_i, q_j)) \cap \text{inv}(q_i) \neq \varnothing$,
$h_{\text{jump}} : \mathcal{T} \times \mathcal{X} \mapsto \mathcal{X}$ is a jump function returning the next continuous state after a transition.

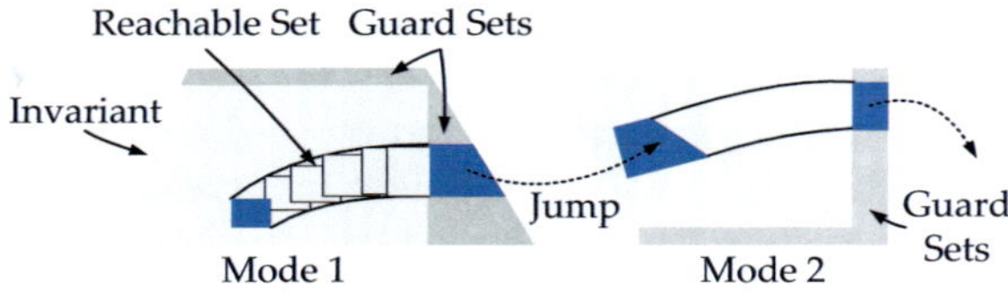

Figure 3.6   Visualization of reachable sets of a hybrid automaton.

The input u is assumed to be locally Lipschitz continuous[2] and the input sets $\mathcal{U}_q$ depend on the mode q. A different set of parameters can be chosen for each mode q. The jump function h_{jump} is restricted to a linear map.

Hybrid systems are well known in process engineering, especially for supervision, e.g., of a chemical process. The supervision task usually has qualitative aims like the question whether the temperature in a reactor is within a safe range. Therefore, one abstracts the physical, continuous-variable system by quantization of all continuous-variable signals. However, it might be difficult to solve the supervision task on the qualitative level due to the mixed continuous and discrete nature of the quantized system, since differential equations have to be solved under several inequality constraints induced by the quantizers.

Stochastic automatons represent the system in a purely discrete description that is abstracted from the quantized system. They enable quantitative performance measures in presence of uncertainty. The stochastic automaton is described by a set of transition probabilities instead of differential equations and inequality constraints. If the system shows a non-linear behavior, methods relying on steadiness or on a Lipschitz constraint can be applied. Completeness of the model is a crucial requirement for the stochastic automaton to allow the transferability of its results to the supervision result for the quantized system [72]. A stochastic automaton will be applied to predict potential future states of different classes of road users in this work. Several aspects of uncertainty will be taken into account where state sets of the road users will be represented by intervals.

Intervals provide an efficient representation of sets. One can apply interval arithmetics — a technique which can be applied to most standard operations and functions — on set representations in form of multi-

---

[2] A function is Lipschitz continuous if it is differentiable everywhere and the absolute value of the derivative is bounded by the defined Lipschitz constant.

dimensional intervals. The $\mathcal{I}$-representation of a multi-dimensional interval describes a set by $\mathcal{I} := [\underline{a}, \overline{a}]$, $\underline{a} \in \mathbb{R}^n$, $\overline{a} \in \mathbb{R}^n$, $\underline{a} < \overline{a}$.

Interval arithmetics are often applied when more accurate techniques for set computations fail or are too time-consuming. Unfortunately, interval arithmetics might provide very conservative solutions resulting in possibly unacceptable over-approximations. For example, formulations using a zonotope[3] representation as in [71] for sets are less sensitive to over-approximation but make the computation much more complex and computationally demanding. Therefore, interval sets are used in this work. Basic operations for interval computations are described in Appendix A.4.

3.5 Reachability Analysis

Reachability analysis is a mean to determine a set of states reachable by a system $f(\mathbf{x}(t), \mathbf{u}(t), \boldsymbol{\rho}(t))$ that depends on time t if it starts from a bounded initial set of states $\mathbf{x}(0) \in \mathcal{X}^0 \subset \mathbb{R}^n$ (e.g., in form of an interval set) with a parameter vector $\boldsymbol{\rho} \in \mathcal{P} \subset \mathbb{R}^p$ and based on a set of input trajectories $\mathbf{u}([0, \tau]) \in \mathcal{U} \subset \mathbb{R}^m$. $\mathbf{u}([0, \tau])$ denotes the union of inputs $\bigcup_{t \in [0,\tau]} \mathbf{u}(t)$ within the time interval $[0, \tau]$. One can compute reachable sets for points in time and for time intervals. In this work, reachable sets are computed for the speeds and the positions of the road users to predict their future states. The sets are represented by intervals. They take into account uncertainty in the estimated state and in the prediction of the road user's behavior.

Reachable sets for time points $\mathcal{R}(\tau)$ correspond to the union of all possible system states at $t = \tau$:

$$\mathcal{R}(\tau) = \left\{ \mathbf{x}(\tau) = \int_0^\tau f(\mathbf{x}(t), \mathbf{u}(t), \boldsymbol{\rho}(t)) \mathrm{d}t \right\}, \tag{3.70}$$

The union of reachable sets at time points within the time interval $t \in [0, \tau]$ results in the reachable set of a time interval $\mathcal{R}([0, \tau])$:

$$\mathcal{R}([0, \tau]) = \bigcup_{t \in [0,\tau]} \mathcal{R}(t). \tag{3.71}$$

---

[3]A zonotope is a polytope which can be obtained as the Minkowski sum of finitely many closed line segments in $\mathbb{R}^n$. Zonotopes are centrally symmetric, compact, convex sets [73].

Exact continuous reachable sets and discrete reachable sets are required for the application in a hybrid automaton. If the continuous reachable set hits certain guard sets, new discrete states may be reached. The computation of a continuous reachable set starts based on the initial state $\mathcal{X}^0$ from its initial mode q^0. One determines the reachable set for time sub-intervals $\mathcal{R}([(k-1)\tau, k\tau])$ and ascertains that the reachable set does not leave the invariant. If the set left the invariant at time point t^{inv}, one would check for the intersection $\mathcal{R}_j^{\text{int}}$ of hit guard sets with the continuous reachable set. Each guard set is associated with a transition (q_i, q_j) determining the next mode. The jump function h_{jump} maps the intersection set $\mathcal{R}_j^{\text{int}}$ to the new mode $\mathcal{R}_j^+$. Although the reachable set may hit several guard sets within one mode, only the evolution of one future mode can be computed at a time. Therefore, one writes the future modes q_j, the reachable sets after the jump $\mathcal{R}_j^+$, and the minimum time for enabling transitions to a list. Then, one works through the list from top to bottom and computes the corresponding reachable sets where all sets and states of a data structure are interpreted as new initial states.

Stochastic differential equations serve the description of continuous stochastic systems where a deterministic drift term and a stochastic diffusion term are used to obtain the derivative of random variables [74,75]. Stochastic automatons with probability-based transitions from one discrete state to the other are able to model discrete stochastic systems where the probability of the transition may depend on a finite set of discrete inputs [70]. A stochastic reachable set of a continuous system for a point in time is used as a synonym for the probability density function of the state in this work, where the definition is adopted from [71], so that the stochastic reachable set for a time interval $t \in [0, \tau]$ can be expressed correspondingly by integration over time

$$
f_x(\mathbf{x}, [0, \tau]) = \int_0^\tau f_x(\mathbf{x}, t) f_t(t) \mathrm{d}t, \quad f_t(t) = \begin{cases} 1/\tau, & \text{for } t \in [0, \tau], \\ 0, & \text{otherwise.} \end{cases} \tag{3.72}
$$

where $f_x(\mathbf{x}, [0, \tau])$ represents the probability density function of the random state vector $\mathbf{x}$ and t is a random variable which is uniformly distributed within the time interval $[0, \tau]$. A conversion of stochastic reachable sets to reachable sets enables the application of stochastic reachable sets in combination with hybrid systems. This reachable set is the subset of the stochastic reachable set with probability values that are non-zero.

3.6 Markov Chain Abstraction

The Markov chain is a very popular example for a stochastic automaton. Markov processes are stochastic processes where the future is conditionally independent from the past history, given the current state. Stationarity is the property of a system to maintain its dynamic behavior invariant from time shifts. If a stochastic process is defined over a finite or countable set, one refers to that discrete-state process as chain. The combination of both properties leads to a Markov chain:

$$p(x_{k+1}|x_k, x_{k-1}, \ldots, x_0) = p(x_{k+1}|x_k) \ \forall 0 \le k. \tag{3.73}$$

State transitions may only occur at time instants $0, \ldots, k$ in case of a discrete-time Markov chain. The Markov property is memoryless, so that all past state information is irrelevant and no state age memory is required. The probability of a transition to a new state depends only on the current state value in a semi-Markov process and state transitions may occur at any time then.

Markov chain abstraction is a method that computes a probability distribution instead of using the original system dynamics. The Markov chain abstraction can be applied to continuous and to hybrid systems. It has to be created in a way that it represents the original system's behavior with sufficient accuracy. The state space of the continuous system $\mathbb{R}^n$ has to be discretized for abstraction first, since Markov chains are stochastic systems with a discrete state space. The continuous state space is sampled to single cells and combined with the discrete modes resulting in one discrete state space. Following, the number of Markov chain states depends on the product of the number of cells of the continuous state space N_{cont} and the number of modes (discrete states) N_{dis}, so that the Markov chain has up to $N_x = N_{\text{cont}} \cdot N_{\text{dis}}$ states. Therefore, the approach is only reasonable for systems with less than three to five state variables. The number of states for the Markov chain is smaller if there is a unique map of the continuous to the discrete state of the hybrid system which is the case when the invariants of a hybrid system do not intersect. Then one only needs to map the continuous state space of the hybrid system to the discrete state space of the Markov chain. The formal definition of a Markov chain is adapted from [70]:

Definition 3.3 *(Discrete-Time Markov Chain):*
  A discrete Markov chain $\mathrm{MC} = (\mathcal{Q}, \mathbf{p}^0, \mathbf{\Psi})$ *consists of*

- $\mathcal{Q} \subset \mathbb{N}^+$: *the countable set of modes,*

- $p_i^0 = P(\mathbf{x}_0 = i)$: *the initial probability with random state* $\mathbf{x} : \Omega \to Q$, *where* $P(\cdot)$ *is an operator determining the event probability and* Ω *the set of elementary events,*

- $\Psi_{ij} = P(\mathbf{x}_{k+1} = i | \mathbf{x}_k = j)$: *the transition matrix enabling the mapping* $\mathbf{p}_{k+1} = \Psi \mathbf{p}_k$.

The Markov chain is updated after each time increment $\tau \in \mathbb{R}^+$ according to the transition matrix Ψ, so that τ states the relation between the discrete time step k (written in the index) and the continuous time t leading to the continuous time $t_k = k \cdot \tau$ at time step k.

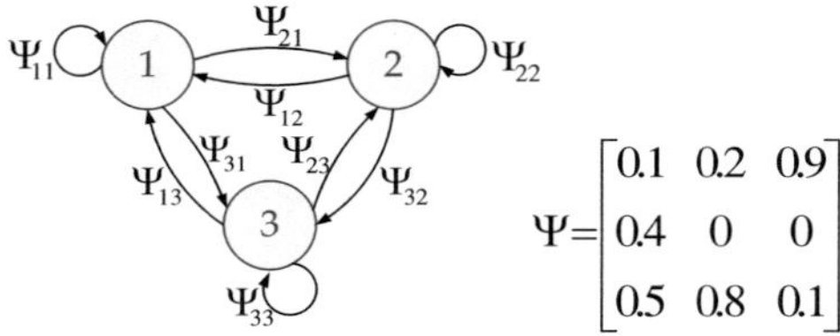

$$\Psi = \begin{bmatrix} 0.1 & 0.2 & 0.9 \\ 0.4 & 0 & 0 \\ 0.5 & 0.8 & 0.1 \end{bmatrix}$$

Figure 3.7 Exemplary Markov chain with three states and transition matrix Ψ.

Figure 3.7 depicts an example for a Markov chain with three states represented by nodes. The arrows represent the transition probabilities Ψ_{ij} from state j to state i ($1 \leq i, j \leq 3$).

The Markov chain is created by sampling the state space of the original, continuous system into cells representing discrete states. Afterward the transition probabilities from one cell to the another are determined and saved in the transition matrix of the Markov chain.

The continuous state space $\mathcal{X} \subset \mathbb{R}^n$ and the input set $\mathcal{U} \subset \mathbb{R}^m$ are discretized to hyper-rectangles of equal size, meaning that each coordinate of $\mathcal{X}$ and $\mathcal{U}$ is fragmented into equidistant intervals. Thus, an n-dimensional interval $\overline{\mathbf{x}}_i =]\underline{\mathbf{x}}_i, \overline{\mathbf{x}}_i] = \mathcal{X}_i$ describes the cell with index i where $\underline{\mathbf{x}}_i, \overline{\mathbf{x}}_i \in \mathbb{R}^n$. Figure 3.8 shows a visualization of the discretization of the state space $\mathcal{X} \subset \mathbb{R}^2$ for the two-dimensional case.

Correspondingly, an m-dimensional interval $\overline{\mathbf{u}}^\alpha =]\underline{\mathbf{u}}^\alpha, \overline{\mathbf{u}}^\alpha] = \mathcal{U}^\alpha$ represents the input cell with cell index α, where $\underline{\mathbf{u}}^\alpha, \overline{\mathbf{u}}^\alpha \in \mathbb{R}^m$. The state space $\mathcal{X}$ is a subset of $\mathbb{R}^n$. The remaining subset in that space $\mathbb{R}^n \setminus \mathcal{X}$ is called the outside cell and has assigned the index 0. The outside cell is elementary when the transition probabilities of the Markov chain shall be calculated.

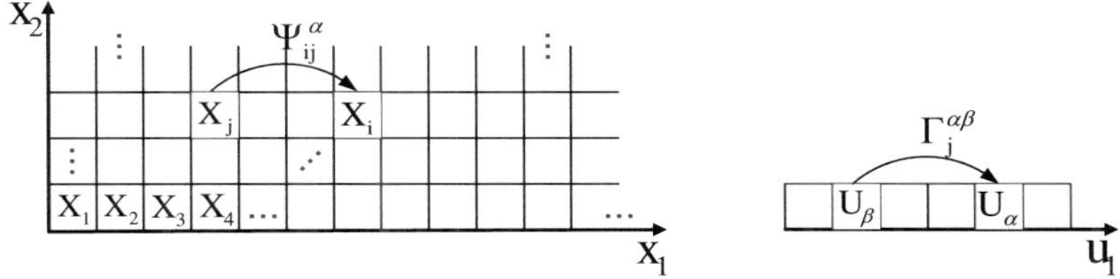

Figure 3.8 Visualization of the discretization of the state space $\mathcal{X} \subset \mathbb{R}^2$ for the two-dimensional case and discretization of the input space $\mathcal{U} \subset \mathbb{R}^1$.

3.6.1 Computation of Transition Probabilities

The continuous dynamics of the original system serve the computation of the transition probabilities Ψ_{ij}. One has to determine the transition probabilities for each discrete input value α and state pair (i, j), so that:

$$\Psi_{ij}^{\alpha} = P(\mathbf{x}_{k+1} = i, \mathbf{y}_k = \alpha | \mathbf{x}_k = j, \mathbf{y}_k = \alpha). \tag{3.74}$$

In this work, the transition probabilities are computed using a large number of simulation runs. Therefore, a final set of initial states is generated from a pre-defined grid on the initial cell $\mathcal{X}_j$. Furthermore, the input cell $\mathcal{U}^{\alpha}$ is discretized to obtain a final set of input values $\mathbf{u}([0, \tau])$. These input values are constant throughout the time interval $[0, \tau]$. The input trajectory $\hat{\mathbf{u}}(t)$ and the parameter vector ρ are based on a corresponding probability distribution and defined input dynamics.

One runs simulations according to the system dynamics $f(\mathbf{x}(t), \mathbf{u}(t), \hat{\mathbf{u}}(t), \rho)$ of the time interval $[0, \tau]$ with combinations of all initial states with all inputs $\mathbf{u}([0, \tau])$. Here, sampling of the initial state cells $\mathcal{X}_j$ and the input cells $\mathcal{U}^{\alpha}$ from uniform grids is applied, but one could also use simulations generated from random sampling (Monte Carlo simulation).

One starts with $N_j^{\mathrm{sim},\alpha}$ simulations in cell $\mathcal{X}_j$ with input $\mathbf{u} \in \mathcal{U}^{\alpha}$. $N_{i,j}^{\mathrm{sim},\alpha}$ is the number of these simulations reaching cell $\mathcal{X}_i$ after time τ, so that the ratio of both numbers leads to the transition probabilities:

$$\Psi_{ij}^{\alpha}(\tau) = \frac{N_{i,j}^{\mathrm{sim},\alpha}}{N_j^{\mathrm{sim},\alpha}}. \tag{3.75}$$

The computation of the transition probabilities for a bounded set of $\tilde{n}$ equidistant intermediate points of time $\tilde{t}_0, \tilde{t}_1, \ldots, \tilde{t}_n \in [0, \tau]$ enables the approximation of the probability that another state is reached within the time interval $[0, \tau]$. The arithmetic mean of the intermediate transition probabilities determines the transition probability for the time interval

$$\Psi_{ij}^{\alpha}([0, \tau]) = \frac{1}{\tilde{n}} \sum_{k=1}^{\tilde{n}} \Psi_{ij}^{\alpha}(\tilde{t}_k). \tag{3.76}$$

Unfortunately, the computation of transition probabilities using Monte Carlo simulation or sample-based simulation does not provide a complete model due to the limited number of simulation runs. The probabilities of some transitions might be set to 0 even though they are non-zero, but there was no simulation for the corresponding state transition. However, the Monte Carlo approach approximates the exact solution quite exactly when the number of simulation runs tends to infinity. Furthermore, the Monte Carlo approach enables the utilization of different noise distribution models for additional inputs, such as Gaussian white noise.

If the continuous dynamics fulfill the Lipschitz continuity, a system can be numerically simulated making the presented approach applicable to all continuous and hybrid systems. One has to ascertain a proper assignment of probabilities to possible executions in hybrid systems with non-deterministic behavior, for example if transitions are enabled but not enforced. As the mode of a continuous state within a cell is uncertain for originally deterministic systems, both the abstraction of deterministic systems and stochastic systems result in a discrete stochastic system.

3.6.2 Markov Chain Update

A Markov chain update means the projection of one state distribution to the next state distribution using the transition matrix. It can be performed for input α and time points k by multiplication of the transition matrix Ψ^{α} with the probability vector $\mathbf{p}_k$. The input $\mathbf{u}$ has to stay within the input cell $\mathcal{U}^{\alpha}$ for the time interval $[t_k, t_{k+1}]$ $(t_k = k \cdot \tau)$, which is denoted by index $[k, k+1]$ in the following, whereas the input $\hat{\mathbf{u}}(t)$ is not considered any longer, as it is already incorporated within the state transition probabilities of the Markov chain. The input $\mathbf{u}$ may change its value for the next time interval. The probability vector $\mathbf{p}_k$ at the beginning of the time interval serves the determination of the probability vector of the time interval $\mathbf{p}_{[k,k+1]}$ as auxiliary term, so that the update is performed according to

$$\begin{aligned}
\mathbf{p}_{k+1} &= \mathbf{\Psi}^{\alpha}(\tau)\mathbf{p}_k\,, & (3.77)\\
\mathbf{P}_{[k,k+1]} &= \mathbf{\Psi}^{\alpha}([0,\tau])\mathbf{p}_k\,. & (3.78)
\end{aligned}$$

The iterative multiplication of the probability distributions with the transition matrices evokes an additional error, as the probability distribution within a cell is treated as if it was replaced by a uniform distribution in the next time step [71]. Smaller discretization steps reduce this error but increase the computational effort.

3.6.3 Markov Chain Update with Uncertain Input

If besides the state of a system, its inputs are uncertain as well, the Markov chain's update requires the introduction of a conditional probability $p^{\alpha|i}$ ($P(\mathbf{y} = \alpha|\mathbf{x} = i)$). The sum over all values of α is the total probability of the state $p_i = \sum_\alpha p_i^\alpha$. The joint probability of the state and the input is given by

$$p_i^\alpha = p^{\alpha|i} \cdot p_i. \tag{3.79}$$

The Markov chain has to be updated for all possible values of α. One obtains the update of the joint probability vector $\mathbf{p}^\alpha$ containing all possible state values for a fixed input α using

$$\begin{aligned}
\mathbf{p}_{k+1}^\alpha &= \mathbf{\Psi}^{\alpha}(\tau)\mathbf{p}_k^\alpha, & (3.80)\\
\mathbf{p}_{[k,k+1]}^\alpha &= \mathbf{\Psi}^{\alpha}([0,\tau])\mathbf{p}_k^\alpha. & (3.81)
\end{aligned}$$

The conditional input probabilities $p^{\alpha|i}$ are updated immediately at time points k to enable the modification of the input probabilities $\mathbf{p}^\alpha$. Therefore, one requires the input transition matrix $\Gamma_{i,k}$ that depends on the current state and the time. It describes the probability of changing the input from input β to input α ($\Gamma_{i,k}^{\alpha\beta}$). Thus, the conditional input probabilities $p^{\alpha|i}$ correspond to

$$p_k^{\alpha|i'} = \sum_\beta \Gamma_{i,k}^{\alpha\beta} \cdot p_k^{\beta|i}. \tag{3.82}$$

If one multiplies Equation 3.82 with $p_{i,k}$ and uses Equation 3.79, it becomes clear that one can update the joint probabilities p_i^α instead of $p^{\alpha|i}$, since the state probability p_i does not change immediately, and therefore:

$$p_{i,k}^{\alpha\,'} = \sum_\beta \Gamma_{i,k}^{\alpha\beta} \cdot p_{i,k}^{\beta}. \tag{3.83}$$

As presented in [71], the joint probabilities p_i^α are combined to a new probability vector $\tilde{\mathbf{p}}^\mathrm{T} = [p_1^1 p_1^2 \dots p_1^{N_u} p_2^1 \, p_2^2 \dots p_2^{N_u} \; p_3^1 \dots p_{N_x}^{N_u}]$ for a simpler notation and an elegant combination of the state transition values Ψ_{ij}^α with the input transition values $\Gamma_i^{\alpha\beta}$, where the values N_u and N_x relate to the number of discrete inputs and states. Consequently, also the corresponding state and input transition values have to be rearranged.

If the dimension of the resulting matrix becomes large and there are only a few non-zero entries meaning that the resulting matrices are very sparse, special algorithms designed for the multiplication of sparse matrices [76] can accelerate the computation. The big advantage of Markov chain abstraction lies in the fact that the computationally expensive part can be accomplished offline ahead when the computation time is almost unlimited. Probability distributions can be computed efficiently during online operation of a system.

This chapter explained the existing methods that have been used in this thesis. Object tracking is performed using two approaches for data association. The concrete implementation for pedestrian tracking with radars and monocular camera will be described in Chapter 4. The situation evaluation module is described in Chapter 5. It uses a Bayesian classifier to distinguish between different maneuver classes. Moreover, stochastic reachable sets are computed using Markov chain abstraction to predict the uncertain future states of the road users.

4 Implementation of two Pedestrian Tracking Approaches

This chapter is dedicated to the description of the system architecture and the implementation of the pedestrian tracking approaches using radars and monocular camera. The first section provides an introduction to existing approaches for pedestrian tracking, fusion of monocular camera and radar as well as to tracking across sensory blind regions. The second section provides information about the hardware and the system architecture, while the further sections introduce the used state model for pedestrians and the ego vehicle as well as the used approach for sensor measurement together with its results. The chapter closes with the description of the applied track management methods.

4.1 Related Work on Sensor Fusion and Pedestrian Tracking

4.1.1 Existing Approaches for Automotive Pedestrian Tracking

Linzmeyer [77] developed a pedestrian detection system using detections from radar and segmented data from thermopile sensors. Object positions are fused separately from the object type using a Kalman filter in combination with simple weighted fusion. Dempster's and Shafer's theory of evidence serves the fusion of the object types.

Fayad and Cherfaoui [78] presented a generic method to fuse data from asynchronous sensors with complementary and supplementary FOVs by tracking detected objects in a commune space. They use standard Kalman filters with nearest neighbor data association to filter pedestrian states. The tracking confidence is calculated and updated based on the score of Sittler [79] using a likelihood ratio, while detection and recognition confidences are updated using basic belief assignment. In experiments, they only use a lidar sensor and do not fuse multi-sensor data nor validate their results. Furthermore, they perform the fusion in a commune sensor space which is not adequate when using different sensor types and accurate measurement covariances shall be provided.

Meuter et al. [80] use an unscented Kalman filter for pedestrian tracking from a moving host using a monocular camera. Their approach does not provide any object existence information and it requires the generation of a region of interest using the 'inverse perspective matching profile', which reduces the computational demand but impairs the detection performance of the algorithm.

Gate et al. [81] use a Kalman filter LNN to track potential pedestrian detections of a laser scanner. The resulting tracks generate regions of interest to classify pedestrians in a video image. The probability that the final tracks result from pedestrians is computed based on detection notes, recognition notes and tracking notes. The advantage of the approach is its low computational complexity; however, the decisive role given to the laser scanner represents a drawback.

Chávez-Garcia et al. [82] fuse data from frontal radar and mono-vision on the detection level before tracking using Dempster-Shafer occupancy grids. Four Kalman filters handle the motion models and the most probable branch is selected as most probable trajectory (Interacting Multiple Model). This approach is not dedicated for asynchronous sensors, has a low detection performance and does not provide existence information about the tracked objects.

Westenhofen et al. [83] generate regions of interest by using a transponder system and detect pedestrians in video images based on HOG features. An EKF tracks the pedestrians (constant turn rate and constant velocity assumption).

Lamard et al. [84] dealt with occlusions in multiple target tracking by modifying the sensor detection probability map in a Multiple Hypothesis Tracker (MHT) and a Cardinalized Probability Hypothesis Density Filter (CPHD), respectively. They presented results using measurements from a camera. In contrast to Reuter et al. [85] who used a sequential Monte Carlo multi-target Bayes (SMC-MTB) filter based on FISST for pedestrian tracking, Lamard et al. modeled the occlusion probability not as a binary value and took into account the uncertainty of the targets state. The presented approach provides a target PoE that is based on the hypothesis probability which uses the track score. However, it does not exploit the inference probability from the measurements for the PoE nor has the approach been presented using detections from radar and monocular camera.

None of the enumerated approaches tracks pedestrians using measurements from radar and monocular camera and provides a reliable PoE. Moreover, only a few authors presented their tracking results using real measurement data. If the other author's approaches provide a PoE, it does not influence the state update or does not exploit the sensory inference probability. There is no approach that uses a JIPDA filter for pedestrian tracking and sensor data fusion with monocular camera and radars. Therefore, a generic approach for pedestrian tracking that fuses data from radars and monocular camera and provides a reliable PoE has been developed in this work.

4.1.2 Related Work on Sensor Fusion using Probabilistic Data Association

The JIPDA filter was published in 2002 for the first time by Musicki [86, 87]. It has originally been designed for aerospace surveillance and is strongly related to other probabilistic approaches for data association, like the PDA [88] or its multi-object variant JPDA [58], which is a sub-optimal single-stage approximation to the optimal Bayesian filter. Restricted extensions of the JPDA allow the formation of new tracks [89]. The JIPDA filter is a sequential tracker in which the associations between several known targets and the latest observations are made sequentially. The state estimation itself is based on an Kalman filter.

Mählisch [60] suggested an implementation of the JIPDA filter to track vehicles, while Munz [65] combined the JIPDA filter with Dempster's and Shafer's theory of evidence to model sensory existence evidence using a monocular camera and a lidar for longitudinal vehicle tracking. Thereby, he enhanced the detection performance of the fusion system. Furthermore, he extended the approach to a generic sensor-independent fusion framework including several improvements in order to guarantee real-time performance of the JIPDA filter — such as a dynamical reduction of gate sizes leading to a limited number of possible combinations and thus, to a bounded computation time. Excluding certain dependencies of association events [90] leads to a significantly simplified computation rule for the association probabilities, which is utilized in the cheap Joint Probabilistic Data Association (cJPDA) filter [91]. This method was applied to vehicle environment perception based on video data in [92]. A more exact approach has been presented in [93], but it assumes a constant detection rate near 1. This is not applicable if the state-dependent probabilities of detection should be accounted for.

Horridge, Maskell et al. [94, 95] circumvent the combinatorial explosion by representing the structure of the target hypotheses in a kind of net. The hypothesis tree consists of repeated subtrees which only need to be computed once. The method exploits the redundancy in an ordered list of objects used to describe the problem. They process the objects in a tree structure exploiting conditional independence between subsets of the objects. The disadvantage of this method is that the complexity strongly depends on the processing order of the objects and the maximum computation time cannot be guaranteed.

Under-segmentation refers to a situation where only one measurement is provided from multiple objects. It may be caused by the limited resolution of a sensor, e.g., poor angular resolution in case of a radar sensor. Measurements from objects located close to each other merge. Zuther et al. have studied this effect in [96]. The opposite effect is called over-segmentation, where several measurements are provided from one object. The effect is particularly relevant when using data from laser scanners. The assumption that one object causes one measurement has to be rejected then. An approach where a set of measurements may be assigned to one object is proposed as multiple association JIPDA in [65]. The last problem is circumvent by merging two measurements to one target if they are expected to result from the same pedestrian.

None of the known approaches utilizes radars and a monocular camera, nor is it specifically designed for pedestrian tracking. In contrast to the work of Munz [65], occluded objects and objects in the close, right sensory blind region are defined as existing objects in this work. Pedestrian tracking with radars and monocular camera from a moving commercial vehicles implies some special challenges, e.g., since the vehicle's cabin strongly vacillates, so that the dynamic camera rotation angles strongly deviate from the stationary ones effecting the camera-based distance estimation. Moreover, no measurement results for the corresponding sensor types have been presented by other authors. Usually, spatially constant sensor noise and constant detection probabilities (except for the outer parts of the FOV) are assumed. Here, the sensors are measured under laboratory conditions and in real-world urban scenarios. The JIPDA filter obtains values dependent on the state and measurement from a look-up-table then.

4.1.3 Existing Approaches for Tracking across Sensory Blind Regions

Different approaches for blind region tracking have been proposed for surveillance, e.g., of public areas or office buildings, using stationary cameras with distinct FOVs. Some assume a known camera topology [97], while others estimate it from the data, e.g., by temporal correlation of exit and entry events [98, 99]. Tracking is performed on each camera using a Kalman filter [100] or a particle filter [101] before the information is shared with the other cameras or before the object information is integrated into a common ground plane tracker, e.g., by intersecting the targets' principal axis [101]. Known transition times and probabilities between the FOVs support the association process [97, 102–104] as well as the object's appearance information [105].

The appearance model may contain a Gaussian distribution for the change of object shape [97, 100, 106] or (color) brightness transfer functions [107, 108], e.g., for (fuzzy) histogram matching [109]. Black et al. [102] use a camera network model to determine the regions where the object is most likely to reappear when an object is terminated within an exit region. The object handover region models consistency of a linked entry and exit region along with the expected transition time between each region [109]. Javed [104] obtain correspondence among cameras by assuming conformity in the transversed paths of people and cars. Moreover, Loke [106], Rahimi [110] and Wang [105] utilize target's dynamics to compensate for the lack of overlap between the cameras' FOVs.

A common characteristic of [97, 100, 103, 107–109] is that they require some form of supervision, or rely on known object correspondences in training data between non-overlapping views, while [98, 99, 102, 106, 110] work unsupervised and operate in correspondence-free manner, which increases the robustness of the algorithm.

Many of the upper approaches provide good results and are adequate for surveillance of areas which the algorithms have been trained for, but the system is moving in this application. Even if the vehicle stops at a red light, the area that has to be observed is different from that around the vehicle of another stop. The trajectories of the objects that walk around the vehicle change as well, so that information like matching trajectory fragments before and after the blind region cannot be used.

A camera is only available for one of the two FOVs here, meaning that no shape matching or appearance model is applicable. The only information that can be used is the matching of the state information (conformity of the object motion) and the measurement properties which indicate that the detection represents a pedestrian. A matching between the reflection amplitude of one radar sensor or the RCS value of the other one cannot be exploited, since the reflection amplitudes of the blind spot radar are distributed rather randomly for all classes of objects than dependent on the object appearance.

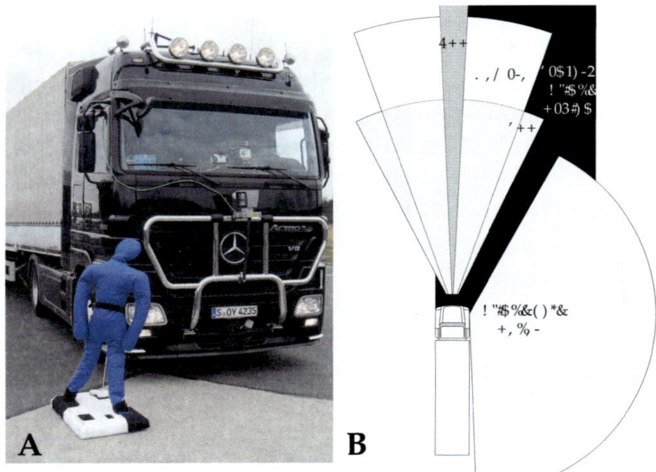

Figure 4.1 A: Testing truck with mounted sensors for frontal environment perception (camera, radar, laser scanner); B: Visualization of the sensory FOVs, where LRR stands for long range radar and SRR for short range radar.

4.2 System Description

The test vehicle that has been used for data recording and system integration during the work on this thesis is shown in Figure 4.1 A. The camera that is used for pedestrian detection is located directly behind the windshield and a few centimeters above the breast. The short range radar sensor (SRR) and the long range radar sensor (LRR) are combined within one housing and their antennas are integrated on the same rotating drum allowing to switch between modes for long range and short range detection. The position of the used radar sensor system for frontal environment perception is below the license plate. A blind spot radar sensor (BSR) monitors a region on the right side of the vehicle and is mounted on the first lower step tread. An illustration of the sensory FOVs is given in Figure 4.1 B. A laser scanner is used as reference sensor to provide ground truth data. It is either mounted on the bull bar or on the upper step tread on the right side of the truck.

The following subsection describes the utilized sensors as well as how pedestrians are detected and classified by these sensors.

4.2.1 Camera

A monocular camera serves the optical detection and classification of pedestrians. It is mounted in a height of 2.47 m with 0.12 m lateral offset to the left behind the windshield. The pitch angle to the ground is 4.2°. The utilized camera optics provide 35° horizontal angle of view and 23° vertical angle of view with a focal length of about 7 mm on axis. A CMOS image sensor converts the image to 752 horizontal pixels and 480 vertical pixels with a bit-depth of 8 bit leading to a resolution of about 21 px/°. The wavelength range of the sensor is within 410 nm and 720 nm with a sensitivity of 4.8 V/lux-sec. The camera expects 12 V for power supply and has a dynamic range of 110 dB. The image processing can be performed in average on 33.33 frames/s. The exposure time is automatically adapted to the environmental illumination conditions.

The image is sent via low voltage differential signaling (LVDS) to a measurement interface hardware that changes the format and sends the image data via universal serial bus (USB) to a computer where the image processing is performed within a framework that is based on C++. The pixels are interpreted as gray values within the framework. A lens correction algorithm as in [111] aims to compensate the lens distortion using the intrinsic parameters of the camera.

Pedestrian Detection and Classification in the Image

A summary of proposed and evaluated methods for pedestrian detection using visible light cameras can be found, for instance, in the contributions of Gandhi and Trivedi [112] or of Enzweiler and Gavrila [113]. Enzweiler and Gavrila [113] provided an overview of monocular pedestrian detection from methodological and experimental perspectives. They cover the main components of a pedestrian detection system and the underlying models. They consider wavelet-based AdaBoost cascade, HOG with linear SVM, neural networks using local receptive fields, and combined shape-texture detection. Their experiments on 20,000 images with annotated pedestrian locations captured on-board a vehicle driving through urban environment indicate a clear advantage of HOG with linear SVM at higher image resolutions and lower processing speeds, and a superiority of the wavelet-based AdaBoost cascade approach at lower image resolutions and (near) real-time processing speeds. More recent approaches for video-based pedestrian detection have been presented by the same authors in [114, 115], while Goto et al. [116] published a cascade detector with multi-classifiers utilizing a feature interaction descriptor for pedestrian detection and direction estimation.

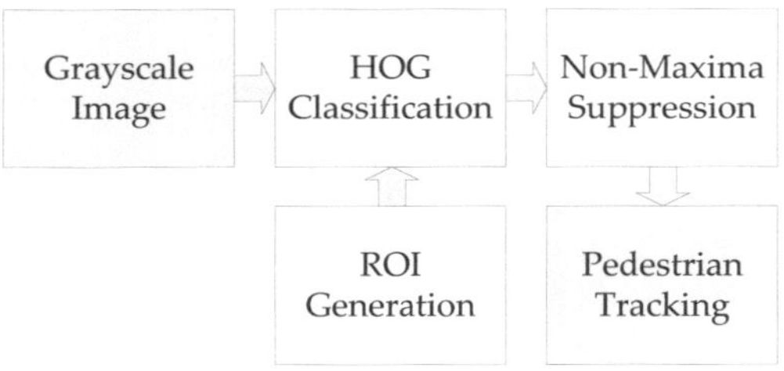

Figure 4.2   Visualization of pedestrian classification approach.

In this work, the camera-based pedestrian detection and classification algorithm is based on the work of histograms of [113, 117, 118] and HOG features with linear SVMs are used. Figure 4.2 visualizes the applied principle. The system utilizes an initial set of regions of interest (ROIs) generated for various detector scales and image locations using a flat-world assumption and ground-plane constraints, which means one selects a box of pixels that represents possible object dimensions at certain distances. The resulting pixel boxes selected as ROI (sub-images) are sent to the classification module that uses features from HOG computations on gray-scale image data [117].

The first classification step returns multiple detector responses at near-identical locations and scales. Therefore, a confidence-based non-maximum suppression (NMS) algorithm performs pairwise box coverage to the detected bounding boxes. Two system detections z_i^I and z_j^I are subject to NMS if their coverage

$$\Xi(z_i^I, z_j^I) = \frac{A(z_i^I \cap z_j^I)}{A(z_i^I \cup z_j^I)} \tag{4.1}$$

is above a certain threshold value (e.g., 0.5), where the coverage $\Xi(z_i^I, z_j^I)$ is the intersection area related to the union area. A bandwidth mean-shift-based mode-seeking strategy determines the position in x-y-scale-space [119]. The final detection's score is the maximum of all scores within the mode. This procedure showed to provide more robust results than applying kernel density to form the final score as done in [120].

The algorithm finally provides a value between -1 and 1 for each object hypothesis to express its confidence whether there is a pedestrian within the sub-image (1) or not (-1). Only hypotheses with a classification result greater than 0 are processed in the JIPDA filter.

4.2.2 Frontal Radar Sensors

The utilized front radar sensors are mounted in the lateral center at the front rear (0.04 m in front of the camera) about 0.39 m above the ground. The lower left corner of Figure 4.3 shows the outer appearance of the frontal radar sensors. The LRR with 17 antennas and the SRR with 15 an-

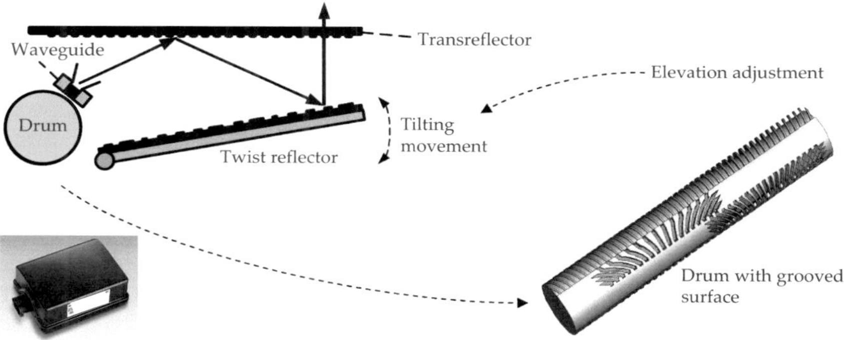

Figure 4.3 Appearance and antenna concept of frontal radar sensors (SRR, LRR), adapted from [11].

tennas are combined within one housing and they use the same electronic control unit (ECU). The mechanically scanning antennas are located on one rotating drum, see Figure 4.3 for the antenna concept. A generator provides a radio frequency signal that propagates along a waveguide. This waveguide is located closely to a drum with a grooved surface building a periodic leaky wave antenna. The electro-magnetic energy is scattered at the grooves of the drum leading to a directed radiation with an angular distribution and amplitude according to the design of the drum surface. Continuous rotation of the drum and different surface properties at different rotation angles enables antenna beam scanning and adjustment of the antenna beam shape in the azimuthal plane of the sensor. Thereby, one obtains different scans with different angular segments and resolutions. The polarizer and the twist reflector build a folded reflector assembly for adjustment in the elevation plane. The wave emitted by the waveguide is reflected at the polarizer and hits the tiltable twist reflector surface shaping the beam in the elevation plane and twisting the polarization plane of the radio frequency wave by $90°$ so that it can pass the polarizer.

The system operates in simultaneous transmit and receive mode, where frequencies between 76 GHz and 77 GHz are used. The maximum angular resolutions are $1°$ for the long range and $4°$ for the short range. Two targets can be resolved as such if they differ in the range more than 2 m or in the range rate more than 5 km/h. The accuracy of the speed measurements is specified to be 0.5 km/h for the LRR and 1 km/h for the SRR. The long range scan provides measurements from an azimuthal aperture angle of $\pm9°$ within a range of $0.25 - 200$ m, while the short range sensor scans the environment from $0.25 - 60$ m within $\pm28°$. The radar sensors have originally been designed for adaptive cruise control applications, so that the detection range is lower for pedestrians (50 m and 35 m) due to weaker reflections and priority to more strongly reflecting targets in the radars' pre-processing unit. The radar accuracies have been measured for pedestrians and the results are presented in Subsection 4.4.2.

The sensor ensemble utilizes a chirped radar modulation scheme (pulse compression radar, chirps of certain shape with different frequencies). The signal-to-noise ratio (SNR) is better compared to a pulse Doppler radar, since a higher amount of radio frequency energy is utilized due to longer duty cycles. Two succeeding fast Fourier transforms (FFT) separate the range and velocity information in the received signals. The transmitted signal has to be designed in such way that the width of the inter-correlated signals is smaller than the width obtained by a standard pulse after matched filtering (e.g., rectangular pulse results in a cardinal sine function). The resulting cycle time corresponds approximately to 66 ms (far and near range scan in one cycle). Figure 4.4 visualizes the main components of the radar sensors in a block diagram.

The sensor arrangement expects some ego vehicle information from the Controller Area Network (CAN), e.g., the vehicle speed. A development interface provides the radar data on the following abstraction levels: peak list, untracked targets (detections with features), tracked objects as well as filtered and tracked objects, whereas the untracked targets (64 for each radar range) are used for pedestrian tracking in this work. A special hardware changes the data format, so that it can be sent via USB to the processing computer. The communication from the sensor to that hardware is performed via an Ethernet connection.

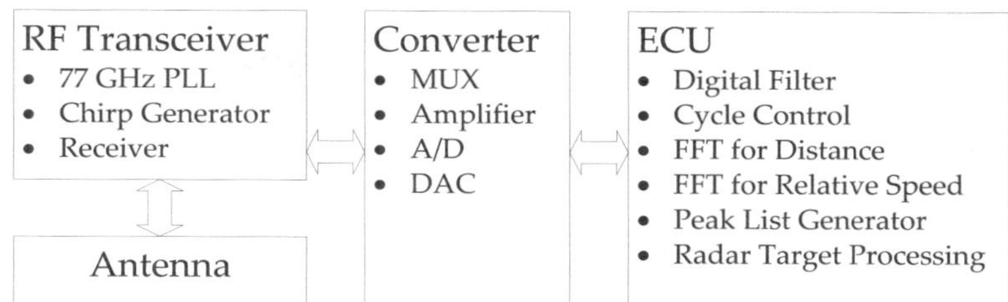

Figure 4.4 Block diagram of frontal radar sensor components: a radio frequency (RF) transceiver with phase-locked loop (PLL) for voltage controlled oscillation to obtain chirps and to receive reflections from the antenna; a converter to multiplex the received, analogue signal (MUX), amplify it and convert it to a digital signal (A/D) as well as to convert digital to analogue signals (DAC) for the reverse direction; an electronic control unit (ECU) for signal processing and control.

Pedestrian Data Filtering from Radar Reflections

The detection performance of the radar may be spatially varying, e.g., due to manufacturing tolerances. The resolution of radar sensors is limited (bandwidth-dependent), especially if a weakly reflecting object (pedestrian) stands close to a bigger and more strongly reflecting object (car). As long as the two objects differ significantly in at least one dimension, the objects may be separated. However, the accuracy of the radar's range and range rate measurement outmatches the camera, since it is almost independent of the cabins pitch angle and of heuristics like a ground plane assumption.

For cars, the positions of reflecting areas on the rear-end can be computed very accurately. However, pedestrians do not reflect only from the body surface but also from inner body parts and due to the radar's mounting position only legs can reflect within short distances. Often only one of the two pedestrian legs is provided as target by the sensors making in hard to track the pedestrian with simple motion models and to set up tracks based on single measurements. The provided Doppler speed is based on the leg measurements as well, so that it varies between standstill and the double pedestrian body speed if the ego vehicle does not move. If both legs are provided — which was usually not the case for the SRR and LRR — merging the measurements to one could help to obtain the pedestrian's body position and speed information. Otherwise, the values of the sensor's measurement covariances have to be increased based on average pedestrian step sizes and the pedestrian speed.

The sensor module computes an RCS value based on the measurements of the SRR and LRR for each detection. It should be around $1\,\text{m}^2$ for complete, adult pedestrians. This measure can be a first indication that the reflecting object is a pedestrian, since objects like cars should provide significantly higher RCS values. There are many other object classes that lead to similar RCS values, e.g., some arrangements of gravel. Moreover, objects with very small RCS values are often suppressed in a dense environment (due to radar design for ACC application), so that not all pedestrian reflections can be obtained then. Furthermore, heavy fog or rain can reduce the assigned pedestrian RCS values, so that they are not detected at all during these weather conditions. The Doppler speed in combination with the ego motion results in another indicator signal for pedestrian filtering. Pedestrians can only move with a certain maximum speed, so that extraordinarily fast targets (e.g., faster than $10\,\text{m/s}$) can be excluded from further computations.

4.2.3 Blind Spot Radar

The radar sensor monitoring the right side of the ego vehicle (blind spot) is referred to as BSR here. It is oriented to the right side and is mounted on the lowest right step tread at the side rear about $0.45\,\text{m}$ above the ground and $0.44\,\text{m}$ longitudinally behind the frontal radar sensors with a lateral offset of $-1.24\,\text{m}$, see Figure 4.5. It uses a single antenna for transmission (single beam monopulse) and four antennas for receiving the echoes. A FMCW measurement principle with so-called star PD wave forms (phase distortion, fast ramping) results in measurements for range and Doppler speed. The carrier frequency is within the range of $76\,\text{GHz}$ to $77\,\text{GHz}$ and the bandwidth for frequency modulation is about $250\,\text{MHz}$. A phase mono-pulse approach leads to angle measurements utilizing four patch antennas on ceramic. The phase shift between antennas with the distance of about $\lambda/2$ (half wave length) is evaluated to obtain an unambiguous measurement for the angle. Another antenna pair with a larger distance (about $3\lambda/2$) is evaluated to increase the accuracy of the measurement. The distances between the single antennas impact the directional characteristic of the sensor. The FOV expands within $\pm 75°$ azimuthal aperture angle and a range of $0.2\,\text{m}$ to $60\,\text{m}$, whereas the sensor is less sensitive at positive angles above $50°$ and in the very near range ($< 1\,\text{m}$). The sensor

Figure 4.5 Mounting positions of blind spot radar and laser scanner at right vehicle side.

is able to discriminate between targets that deviate at least 1 m in range or around 0.5 km/h in speed — parameters that depend on the bandwidth.

Similarly to the SRR's and LRR's RCS value, the BSR provides an amplitude value that could be applied to filter pedestrians, but the evaluation (Section 4.4.3) showed that the signal quality is very poor, so that it is not used.

The BSR operates at a 12 V basis and targets are sent to the computer for tracking via a private, high speed CAN. Here, only two devices are connected to each other and the BSR is the only device that transmits data on the CAN. The message IDs correspond to the 64 target IDs of the sensor. This leads to the special case of constant delay times for transmission over this serial bus. Peak-CAN-adapters convert the CAN data and send it via USB to the computer for further processing.

4.2.4 Laser Scanner

A laser scanner running on a 12 V basis has been used as reference sensor for pedestrian detections of the sensors used for pedestrian tracking. The measurement accuracy of single measurements (reflections) is very high. However, the sensor does not provide complete pedestrian measurements but the distance with the corresponding angle in which laser beams have been reflected by the environment. Thus, the pedestrian data has to be computed from the point cloud which is described after enumeration of the technical sensor data.

The sensor has 64 vertically fixed, eye-safe lasers for scanning the environment mechanically to obtain 360° FOV in the azimuth and $+2°$ to $-24.8°$ in the elevation. Each laser sends 5 ns pulses with a wavelength of 905 nm. While the angular resolution of single measurements in the azimuth is specified to be 0.09°, it should be about 0.4° in the elevation. The detection range for materials with low reflectivity (e.g., pavement) is 50 m and increases with a higher reflectivity (e.g., cars) to more than 120 m, whereas the range accuracy is specified to be below 2 cm. Entire unit spins are performed at a rate of 5 to 15 Hz (300 to 900 rotations per minute). The sensor is able to measure more than $1.33 \cdot 10^6$ reflection points per second and sends its data via Ethernet to the processing computer at a rate of 100 Mbps using the user datagram protocol (UDP). The internal latency of the measurement results is specified to be smaller than 0.05 ms.

Pedestrian Data Filtering from Scanner Data

Figure 4.6 shows the main computation steps to obtain pedestrian detections from the scan data. Ground reflections are removed from the data

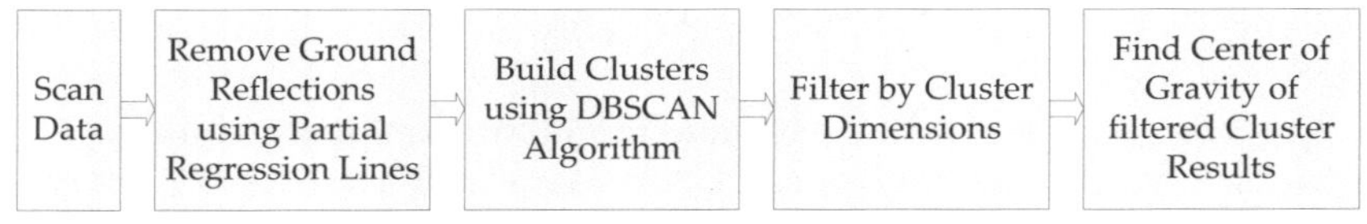

Figure 4.6 Procedure for pedestrian filtering from laser scanner data.

in a first step by estimating ground plane patches of a few square meters. It is not advantageous to estimate only one ground plane since, for example, sidewalks in raised position would evoke aslope ground planes, so that not all ground plane points could be removed on the one side and relevant data would be eliminated on the other side. For estimation of the ground plane patches, the 3D reflection points are reduced to two dimensions by setting their lateral value to zero ($y = 0$) first. Next, one estimates a partial regression line through the resulting points using the least squares method. Appendix A.5 briefly describes the applied least squares method that provides the coefficients of the partial regression line. The application of the **y**-coordinate vector of the data enables the description of the corresponding ground plane patch. All scan points below this patch and those closely above are removed from the data.

The remaining points belong to the objects of the environment, e.g., houses, bushes, cars, street lamps or pedestrians. An algorithm for density-based spatial clustering of applications with noise (DBSCAN) proposed by Ester et al. [121] in 1996 helps to build clusters of scan points that represent these extended objects. The advantage of this cluster algorithm is that it requires only one input parameter as domain knowledge and supports the user by determining it. Furthermore, it discovers clusters with arbitrary shape, which is important for pedestrian detection, and it works very efficiently even under noise. The complexity is $O(N_{\text{points}} \log N_{\text{points}})$, where N_{points} is the number of points. For example, Chameleon [122] is another hierarchical clustering algorithm that provides good cluster results but has a higher complexity $(O(N_{\text{cl}} N_{\text{points}} + N_{\text{points}} \log N_{\text{points}} + N_{\text{cl}}^2 \log N_{\text{cl}})$, where N_{cl} is the number of clusters). It requires the number of clusters as given input which is not available when detecting a variable number of objects in the environment [123]. The shared near neighbor algorithm (SNN) by Ertoz et al. [124] is strongly related to DBSCAN but defines the similarity between points by looking at the number of nearest neighbors that two points share. It is thereby able to detect clusters with different densities, which is not required here, since the density of reflection points from laser scanner shows only little variation in the considered near range.

The DBSCAN algorithm is utilized for several applications in diverse disciplines, for example, for the creation of thematic maps in geographic information systems, for processing queries for proteins with complementary surfaces in molecular biology, or for separating discrete sources from noise of a survey of the sky in astronomy [125]. The key idea of the DBSCAN algorithm is that the neighborhood of a given radius has to contain at least a minimum number of points for each point of a cluster, meaning that the density in the neighborhood has to exceed a certain threshold. The choice of a distance measure between point pairs ($\langle q_s, q_t \rangle$, $s \neq t$) determines the shape of a neighborhood. Here, the Euclidean distance between point pairs is applied as distance measure.

Ester et al. [121] define a threshold distance ϵ for the neighborhood $N_\epsilon(q_s)$ of a point q_s. There has to be another point q_t in the database $\mathcal{D}_{\text{points}}$ that is within the range of ϵ:

$$N_\epsilon(q_s) = \{q_t \in \mathcal{D}_{\text{points}} | \text{dist}(q_s, q_t) \leq \epsilon\}. \tag{4.2}$$

Figure 4.7 visualizes different classes of points that will be defined and their neighborhood regions. If a point q_t has at least $N_{\min}$ points q_s in its ϵ-neighborhood $N_\epsilon(q_t)$, it is a core point ($q_s \in N_\epsilon(q_t) \wedge |N_\epsilon(q_t)| \geq N_{\min}$).

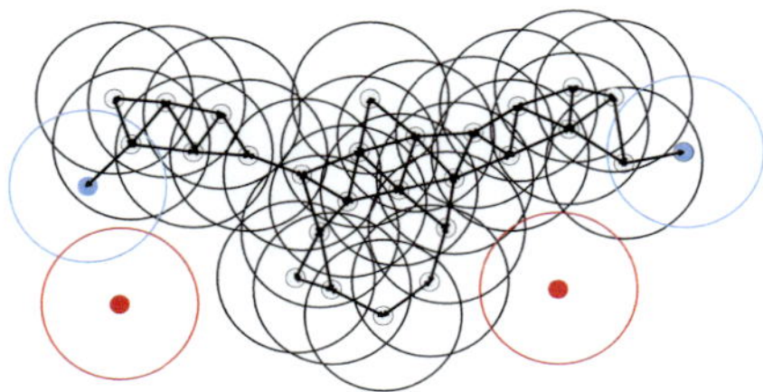

Figure 4.7   Cluster with visualization of DBSCAN with $N_{\mathrm{min}} = 3$, where the two blue points represent border points that are density connected via the gray core points. They are density reachable from the core points. The red points represent noise.

Each point in the ϵ-neighborhood of that core point is directly density reachable from q_t. The relationship of direct density reachability is non-symmetric.

The definition is closely related to density reachability of a point q_s from a point q_t with respect to ϵ and N_{min}. One claims that there is a chain of points $q_1, \ldots, q_n$, $q_1 = q_t$ and $q_n = q_s$, so that subsequent points in that chain are directly density-reachable from q_s. Two border points q_s and q_t of the same cluster $\mathcal{C}_{\mathrm{cl}}$ might not be density reachable from each other. However, there might be a core point q^{core} in the cluster that is density reachable with respect to ϵ and N_{min} from both border points q_s and q_t, so that the border points are density connected.

Finally, the formal definition of a cluster $\mathcal{C}_{\mathrm{cl}}$ with respect to ϵ and N_{min} follows. It is a non-empty subset from the database of points $\mathcal{D}_{\mathrm{points}}$ that fulfills the conditions for maximality and connectivity:

1. $\forall q_s, q_t$: if $q_s \in \mathcal{C}_{\mathrm{cl}} \wedge q_t$ is density-reachable from q_s with respect to ϵ and N_{min}, then $q_t \in \mathcal{C}_{\mathrm{cl}}$.

2. $\forall q_s, q_t \in \mathcal{C}_{\mathrm{cl}}$: q_s is density connected to q_t with respect to ϵ and N_{min}.

Noise can be defined as subset $\mathcal{S}^{\mathrm{noise}}$ of the database $\mathcal{D}_{\mathrm{points}}$. It contains all points that do not belong to any cluster $\mathcal{C}_{\mathrm{cl,id}}$ ($\mathcal{S}^{\mathrm{noise}} = \{q_s \in \mathcal{D}_{\mathrm{points}} | \forall \mathrm{id} : q_s \notin \mathcal{C}_{\mathrm{cl,id}}\}$), see Figure 4.7 for an illustration of different point classes.

The parameters ϵ and N_{min} have to be defined before a cluster can be discovered. Then, one chooses an arbitrary, non-classified point from the database $\mathcal{D}_{\mathrm{points}}$ and sets it as seed point q^{center}. A region query around the seed point q^{center} collects all N_ϵ points lying within the radius ϵ around q^{center} and checks whether q^{center} fulfills the core point condition. In case q^{center} is not a core point, q^{center} is temporarily classified as noise and the algorithm iterates over the remaining non-classified points in the database $\mathcal{D}_{\text{points}}$ until it finds a point fulfilling the core point condition. Next, the algorithm tries to find all points in the database that are density reachable from that core point. If N_ϵ is equal to the minimum number of required neighborhood points $N_{\min}$ or exceeds it, the center point q^{center} is a core point. All points within its ϵ-neighborhood are classified as members of a new cluster $\mathcal{C}_{\text{cl,id}}$ with an unused ID id. Moreover, they are considered as new seed points $q_t^{\text{seed}} \in \mathcal{S}^{\text{seed}}$, $t = 1, \ldots, N_\epsilon$, if they have not already been classified as noise. Next, the original seed point q^{center} is deleted from the seed set $\mathcal{S}^{\text{seed}}$. Each remaining seed point $q_t^{\text{seed}} \in \mathcal{S}^{\text{seed}}$ is considered a new center point, so that further region queries may result in additional seed sets $\mathcal{S}_{q_t}^{\text{seed}}$ if the points q_t^{seed} fulfill the core point condition. Then, all points in the ϵ-neighborhood of the new core points are density reachable and belong to the cluster $\mathcal{C}_{\text{cl,id}}$.

The used center points are removed from the seed sets $\mathcal{S}^{\text{seed}}$ after the region query. The algorithm iterates over the seed sets until they are empty. Then it chooses new non-classified points from the database $\mathcal{D}_{\text{points}}$ and repeats the procedure with new cluster IDs until all points in the database have been classified. Although some points may have been classified as noise initially, they may be located in the ϵ-neighborhood of another point, so that they can be classified as border point of a cluster $\mathcal{C}_{\text{cl,id}}$ later. A flow chart of the DBSCAN algorithm is contained in Appendix A.6.

Spatial access methods, such as R^*-trees [126], support the efficient computation of the ϵ-neighborhood of the points which decreases the complexity of the algorithm.

The parameters ϵ and $N_{\min}$ can be obtained interactively using a sorted k-dist graph [121] for a given k. One iterates over all points of the scan database $\mathcal{D}_{\text{points}}$, computes the distances of each point q_s to all other points within the set of scan points $\mathcal{D}_{\text{points}}$ and sorts the distances in ascending order for each point q_s. One takes the k-th distance value from each list and writes it to another list. This list is then sorted in descending distance order and is plotted. At some point, a kind of valley can be observed and the threshold ϵ is set to the distance value at the 'valley.' $N_{\min}$ is set to the value of k. All points that show higher k-dist values are considered as noise then. The choice for k is 4 in this work, since the 3D data represents a surface instead of a volume, see Figure 4.8.

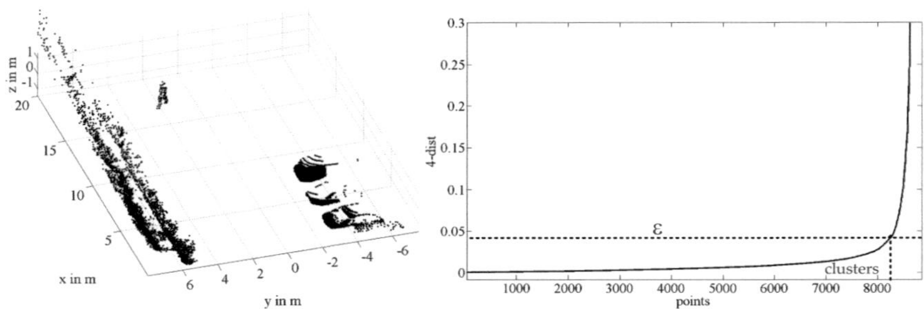

Figure 4.8 4-dist graph to obtain the distance threshold parameter of the DB-SCAN clustering algorithm. The computation was performed based on the points of the clusters on the left side.

Subsequently, the obtained clusters are filtered by their dimension. Pedestrians are not expected to be taller than 2 m or smaller than 1 m. Their lateral expansion is assumed to be within 0.5 m and 1.5 m. With these limitations, only a few clusters remain and the centers of gravity of these clusters are taken as the longitudinal and lateral values of the pedestrian detections. When the scanner is mounted on the step tread on the right side, it can only see reflections from pedestrian legs in the short range, so that the dimensions for filtering are adapted to leg detection in this range and two legs are fused to one pedestrian detection.

The described approach is quite simple but works very well in un-crowded scenes that have been used for measurement of the sensors described above. All pedestrians could be detected and there were no false detections in the scenes. More complex approaches that include the computation of an outline contour of vehicles have been presented by Steinemann [127] using laser measurements or by Kidono [128], who recognized pedestrians using high-definition lidar. It is likely that the latter works more reliably in crowded scenes than the approach presented above considering the false alarm rate but it will hardly provide a better detection rate.

Ego Vehicle

Since the ego vehicle is moving in traffic, the so-called ego motion has to be taken into account during the tracking procedure. The absolute speed v_e of the back axle and the yaw rate $\dot{\Phi}_e$ serve the computation of the ego vehicle's states. The information is received from the vehicle CAN after it has been filtered by an extended Kalman filter in the ECU of the electronic stability program. The speed is computed based on the wheel speeds and the wheel circumference. Mählisch [60] proposed an approach that estimates the ego motion $\hat{\mathbf{x}}_e = (v_e, \dot{\Phi}_e)$ and its covariance matrix $\hat{\mathbf{P}}_e$ based on the measurement's steering wheel angle, wheel speeds of all single wheels and the yaw rate. Unfortunately, no single wheel speeds are available on CAN here, so that this information cannot be exploited. An additional process function f_e models and compensates for the ego motion after the prediction of the objects' states in each cycle. The resulting state uncertainty is taken into account in the updated covariance matrix $\hat{\mathbf{P}}_{k|k-1}$. The correctness of the sensors' distance estimation depends on the correct

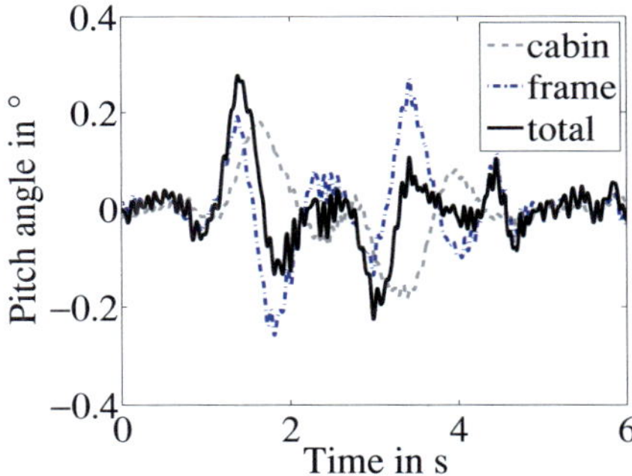

Figure 4.9 Dynamic pitch angle over time when driving over a piece of wood with trailer at 50 km/h without consideration of tire suspension.

estimation of the rotation angles, especially, on the pitch angle of the camera. The rotation angles are dynamic when the ego vehicle is moving due to suspension. The low-frequent part of the pitch angle variation due to acceleration and deceleration can be estimated and compensated online using a linear function that determines the dynamic pitch angle offset dependent on speed and acceleration. The more high-frequent part of the pitch angle variation, e.g., due to driving over rough ground or gully covers, is taken into account in the sensors' measurement covariance matrices $\mathbf{R}$. Tests on various surface types (including driving over a piece of wood) with different speeds have been performed. The spring deflection between the cabin and the frame has been measured at the four corners as well as the spring deflection between the frame and the axles. Thereby, the resulting additional variance of the camera coordinates could be calculated dependent on the pitch angle variance. Figure 4.9 shows a plot of the resulting pitch angle offset over time when driving over a piece of wood with trailer at 50 km/h. The maximum deviation of the pitch angle from the static case increases when the truck does not pull a trailer due to the reduced mass on the rear axle.

4.3 Coordinate Systems and Sensor Calibration

Different mounting positions and orientations of the single sensors require that the data is transformed from the sensors' coordinate systems to one common vehicle coordinate system. Its origin is defined in the middle of the rear axle on the ground. The definition for axis and angle orientation is chosen according to the DIN70000 [129]. Thus, the x-axis points longitudinally to the vehicle front, the y-axis builds the perpendicular, lateral component pointing to the left, while the z-axis points to the top (right-handed coordinate system). The rotations yaw, pitch and roll are defined positive when they are applied counter-clockwise (mathematically positive). Figure 4.10 illustrates the applied coordinate system. The coordinates in the vehicle coordinate system are denoted by x_{VC}, y_{VC} and z_{VC}.

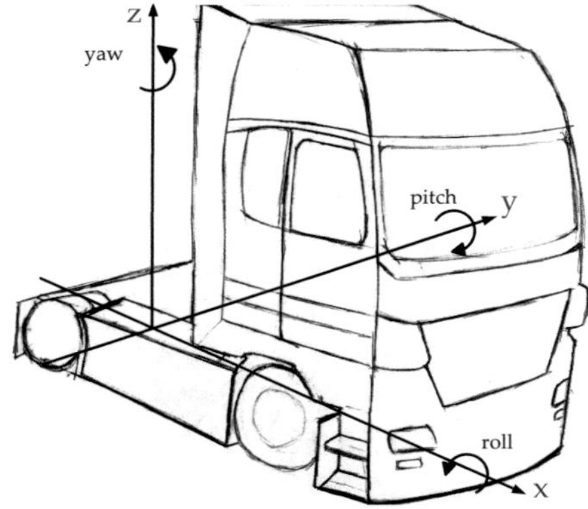

Figure 4.10   Illustration of the applied vehicle coordinate system according to DIN 70000.

Three-dimensional rotation and translation enable the transform of the Cartesian sensor coordinate systems into the Cartesian vehicle coordinate system and vice versa. Utilization of homogeneous coordinates

$\mathbf{x}_H = (x, y, z, 1)^T \in \mathbb{R}^4$ is recommendable. Then, the transform to vehicle coordinates $\mathbf{x}_{H,VC}$ is a single multiplication of the homogeneous sensor coordinates $\mathbf{x}_{H,SC}$ with the transformation matrix

$$\mathbf{T}_{SC2VC} = \begin{pmatrix} \mathbf{R}_{rot} & \mathbf{t} \\ \mathbf{0} & 1 \end{pmatrix} \in \mathbb{R}^{4 \times 4}, \tag{4.3}$$

where

$$\mathbf{R}_{rot} = \mathbf{R}_{rot,x}(\alpha)\,\mathbf{R}_{rot,y}(\beta)\,\mathbf{R}_{rot,z}(\gamma), \tag{4.4}$$

$$\mathbf{R}_{rot,x}(\alpha) = \begin{pmatrix} 1 & 0 & 0 \\ 0 & \cos(\alpha) & -\sin(\alpha) \\ 0 & \sin(\alpha) & \cos(\alpha) \end{pmatrix}, \tag{4.5}$$

$$\mathbf{R}_{rot,y}(\beta) = \begin{pmatrix} \cos(\beta) & 0 & \sin(\beta) \\ 0 & 1 & 0 \\ -\sin(\beta) & 0 & \cos(\beta) \end{pmatrix}, \tag{4.6}$$

$$\mathbf{R}_{rot,z}(\gamma) = \begin{pmatrix} \cos(\gamma) & -\sin(\gamma) & 0 \\ \sin(\gamma) & \cos(\gamma) & 0 \\ 0 & 0 & 1 \end{pmatrix}. \tag{4.7}$$

$\mathbf{R}_{rot} \in \mathbb{R}^{3 \times 3}$ is the rotation matrix depending on the yaw angle α, the pitch angle β and the roll angle γ and the translation vector $\mathbf{t} \in \mathbb{R}^3$, so that

$$\mathbf{x}_{H,VC} = \mathbf{T}_{SC2VC} \cdot \mathbf{x}_{H,SC}. \tag{4.8}$$

The FOV of each sensor is described in the corresponding sensor coordinate system using a value for the range and the horizontal aperture angle. The fusion process takes place in the measurement space of each sensor, so that a measurement function $h(\cdot)$ transforms all state predictions (prior states) to the measurement space. Moreover, track initialization requires a function that transforms single measurements to the state space. The applied measurement functions are provided in Section 4.5.

4.3.1 Spatial Calibration

The spatial calibration determines the rotation angles between the single sensors and between the vehicle coordinate system and the sensor coordinate systems as well as the corresponding translations. Thus, it provides the required parameters of the transformation matrices $\mathbf{T}_{RC2VC}$, $\mathbf{T}_{CC2VC}$ and $\mathbf{T}_{LC2VC}$ transforming radar coordinates, camera coordinates and laser

Figure 4.11   Calibration Target.

scanner coordinates to vehicle coordinates. The relative position and orientation of two sensors, e.g., camera and radar, is retrievable by

$$\mathbf{T}_{RC2CC} = (\mathbf{T}_{CC2VC})^{-1}\mathbf{T}_{RC2VC}. \tag{4.9}$$

A special target is applied for cross calibration in this work, see Figure 4.11. Three quadratic metal plates (1 m × 1 m) are arranged perpendicularly to each other. The target is positioned at different distances to the vehicle in various, relative angles, so that the whole relevant measurement space is covered. The inner side of the target should be oriented to the sensors. Furthermore, the diagonal through the ground plate and the intersection point of all three plates should also intersect with the sensors' x, y-positions. The arrangement of the metal plates serves as corner reflector for the radar, so that the received reflections result from the vertical edge of the target. The laser scanner can detect the same edge, since there are several scan points on the target. If one fits three planes into these points and intersects the resulting vertical planes, the result is the vertical edge. A checkerboard pattern on each metal plate enables the detection of the target by the camera. Before the extrinsic calibration of the camera is performed, the camera will already have undergone an intrinsic calibration, so that the focal length and the principal point of the optical mapping are known. Moreover, an algorithm will have corrected the lens distortion using a procedure that is described in [111]. An image is then taken and an image processing algorithm detects the points where the white and black squares meet by correlation of all rotation possibilities of a scaled checkerboard pattern template at all image positions. One obtains maximum correlation coefficients when the pattern position and orientation matches to one of the three target sides. Thereby, one retrieves the positions of the crossing points in image coordinates [130].

The knowledge that all target plates are perpendicular to each other can be used to set a constraint. The sensor vehicle and the calibration target are located on the same even plane for all target distances. The boundary points of the squares on the vertical plates build parallel rows in defined, known heights over the ground plane, whereas the rows of one plate are perpendicular to those of the other one. Different target positions provide different image coordinates of the boundary points. Fitting a plane through the boundary points enables the computation of the height of the camera mounting position, the roll and the pitch angle.

The longitudinal and lateral position of the camera is measured using standard measurement equipment. Positioning of the calibration target in various distances on the elongation of the vehicle's longitudinal axis enables the computation of the yaw angle within an acceptable accuracy.

Since the position and the orientation of the camera in the vehicle coordinate system are known now and are contained in the transformation matrix $\mathbf{T}_{CC2VC}$, cross calibration to the camera can be applied to compute the extrinsic calibration parameters of all remaining sensors. Sensor positions are measured approximately using standard measurement equipment in a first step. The pitch angle of the radar sensor is neglected. The front radar and the camera shall provide the same lateral and longitudinal value in vehicle coordinates for the vertical edge of the target providing an equation system based on target measurements in different positions up to 45 m. The solution of the equation system using error minimization with the least squares method provides the parameters of the transformation matrix $\mathbf{T}_{RC2CC}$ enabling the computation of the transformation matrix $\mathbf{T}_{RC2VC}$ of the front radar.

In a similar way, one calculates the transformation matrix of the laser scanner mounted at the vehicle front but without negligence of the pitch angle by setting the constraint that the vertical edge of the target which has been detected by the laser scanner has to be perpendicular to the ground plane. Table 4.1 summarizes the resulting positions and orientations of all sensors including the reference sensor (laser scanner). The BSR does not have a common FOV with the front sensors. However, if the laser scanner is mounted on the second step tread of the right side, it detects targets at the front within an angle of about 120° (210° total). A corner reflector on a thin stand is positioned at various distances and angles within the common FOV of the laser scanner and the SRR. The corner reflector serves as target for the SRR, while the laser scanner at the side detects the stand. This enables a cross calibration between SRR and the laser scanner with sufficient accuracy. Next, the BSR is cross calibrated to

Sensor	Position			Orientation		
	x	y	z	roll - α	pitch - β	yaw - γ
Cam	5.28 m	0.12 m	2.47 m	1°	4.2°	0.3°
SRR	5.32 m	0.00 m	0.39 m	0°	0.0°	0°
LRR	5.32 m	0.00 m	0.39 m	0°	0.0°	0°
BSR	4.88 m	-1.24 m	0.45 m	0°	0.0°	−102°
Laser Front	5.49 m	0.00 m	1.70 m	0°	3.0°	0°
Laser Side	4.88 m	-1.44 m	0.70 m	0°	0°	−88°

Table 4.1 Sensor positions and orientations.

the laser scanner, so that the transformation matrix $\mathbf{T}_{\text{BC2VC}}$ is obtained. Table 4.2 shows the obtained detection ranges and corresponding aperture angles.

Sensor	FOV angle	Maximum range	Minimum range
Cam	±17.5°	70 m	10 m
SRR	±28°	35 m	2 m
LRR	±9°	50 m	2 m
BSR	±75°	30 m	2 m
Laser Front	360° (180°)	45 m	7 m
Laser Side	360° (180°)	45 m	2 m

Table 4.2 Sensor FOVs for pedestrian detection.

4.3.2 Temporal Calibration

The state estimation based on motion models and measurement data should refer to the same time point for data association. The timestamp of a measurement has to be known exactly in the prediction step or when using an initialization approach based on several measurement cycles. For example, if the sensors do only provide the position of dynamic obstacles, the speed can be estimated from differentiation of the positions and exact time increments. Therefore, a fusion architecture should provide a global system time with sufficiently high resolution and accuracy. Each sensor measurement should get assigned a corresponding timestamp then and all computations (update and prediction) should be performed based on these. However, the sensors that are available here do not provide timestamps based on a global time, so that they can only get assigned a timestamp when they are received in the processing computer. The sensor-specific delays due to measurement duration (acquisition time), sensor-internal preprocessing and communication have to be determined but underlie stochastic variations. The delays are not perfectly constant, e.g., since the radars and the camera send their data at cycle times that are computed internally and not directly after measurement. However, the constant part of the temporal offset can be determined and compensated.

The fusion must be adapted to the delay of the slowest sensor system. Worst case delays arise when a measurement from the sensor with the shortest latency is received immediately after a measurement from the sensor with the longest latency. Sensor measurements should be processed in the order of measurement to avoid a defective state estimation in the JIPDA filter. If the latencies of the single sensors lie within the same magnitude, the influence of different latencies is small and the measurements can be processed in the order of arrival in the computer and the delay is minimized. Otherwise, it is recommendable to store arriving measurement data in a buffer and to process it in the correct order, so that one measurement cycle from each sensor has been processed before data from the first sensor is processed again. The latency problem can be solved by a time-indexed buffer of observations, state vectors and existence estimations. The buffer size depends on the maximum acceptable observation delay. Westenberger et al. [131] proposed and compared two methods to deal with out-of-sequence measurements in JIPDA for temporally asynchronous measurements. Here, a time-indexed buffer is applied. Measurements from sensors with a higher update rate can be neglected if later measurements arrive before there are measurement cycles of the sensors with longer cycle times.

As described above, two frontal radar sensors (SRR, LRR) with identical latencies, a monocular camera and a radar monitoring the vehicle's right side (BSR) provide the measurements for the fusion process. Their constant-delay offsets (measurement to arrival in computer) have to be determined as parameters for the fusion framework. Huck et al. [132] presented an approach for exact time-stamping of asynchronous measurements in a multi-sensor setup that was based on the work of [133, 134].

They compensated jittering timestamps using filter techniques, proposed a procedure to get timestamps closer to unknown measurement timestamps and determined latencies between measurements from different sensors with a test drive using cross-correlation.

Here, the same procedure has been used. The latency of the camera has been computed to be 42 ms, which does not include the image processing that is done in the computer. The SRR and LRR show a latency of 133 ms. The BSR could have been temporarily mounted at the front of the ego vehicle. The latency of the laser scanner has not been computed but estimated. The procedure described in [132] is not directly applicable here since the laser scanner does not send after whole rotations. The impact of this simplification is negligible, since pedestrians and the ego vehicle move rather slowly and a fixed definition of a pedestrian position induces higher tolerances. It is hard to define a reliable position of a pedestrian, since he changes his shape during motion. The filtered ego motion data that is received via a public vehicle CAN from the ESC-ECU does not have a constant latency either, especially due to the numerous devices that send on the CAN with different priorities and cycle times. However, speed and yaw rate change only slowly in relevant scenarios, so that the resulting inaccuracies are acceptable.

4.4 Sensor Measurement

All Bayesian state estimators require knowledge about the measurement variances and covariances. The JIPDA filter approach for pedestrian tracking expects information about some additional properties like the probability of missing detections $(1 - p_D)$ and the probability of false alarms p_{FP}. The probability of detection p_D and the probability for a false alarm p_{FP} characterize the detection quality of a sensor. The values can be derived from statistical detection failures like missed detections (false negatives) and false alarms (false positives). The JIPDA takes these probabilities into account during the tracking and the fusion process, quantifying the existence of an object x_i via the PoE $p(\exists x_i)_{k|k}$ at time point k. Ghost objects (false alarms) should induce smaller PoEs than real objects.

Different approaches in literature assume that the false alarm probability is a prior known, parametric, spatial distribution. Mostly, a Poisson distribution is assumed, e.g., in [86]. The assumption of constant false alarm probabilities is usually not fulfilled, since the sensor vehicle moves in the environment through various situations. Therefore, the false alarm probability $p_{\mathrm{FP}}(z_j)$ is chosen dependent on the measurement properties here. The attributes of the measurement determine the probability that the measurement is a false alarm. Thus, a sensory inference probability is used to circumvent assumptions about constant false alarm probabilities. Inference probabilities are easy to obtain from classification-based detectors. As described above, a HOG classifier is applied for pedestrian detection with subsequent non-maximum suppression. A ROC curve depending on the resulting confidence values has been computed using labeled ground truth data. Thus, the inference probability of a false alarm is the ROC value that corresponds to the confidence value of a detection after non-maximum suppression. Accordingly, the posterior probability for a true positive is $p_{\mathrm{TP}}(z_j) = 1 - p_{\mathrm{FP}}(z_j)$.

The probability of detection of a sensor p_{D} describes the confidence that an object is detected by the sensor. As the probability of detection often depends on the current state of an object $\mathbf{x}_i$, it should not be modeled uniformly. Usually, it is retrieved from statistics. The probability of detection may depend on different orientations of the objects to the sensor, the weather or the illumination (e.g., motion blur at sunset makes detection performance of the camera worse). Furthermore, sensor-dependent occlusion could be taken into account. For example, a radar usually detects objects that are located on the same angle within a distance of a few meters, whereas the camera usually does not detect the second object then.

Statistical information for the sensors' measurement variances and the other two sensor characteristics $p_{\mathrm{FP}}(z_j)$ and $p_{\mathrm{D}}(x_i)$ requires reference data. The covariance matrices of the sensors' noise have been obtained using pedestrian detections from the laser scanner as ground truth data, whereas it has been manually checked that no false alarms were in the ground truth data. The measurements have been recorded from the stationary sensor vehicle using different pedestrians equipped with various accessories. As the JIPDA requires information about the state-dependent detection performance of the sensors, measurements with other objects that are no pedestrians have been performed as well to trigger false alarms. Here, not only measurements from the stationary sensor vehicle but also from test drives in town and on country roads were included in the data. Moreover, information from manually labeled video sequences that have been recorded on urban streets was utilized as additional ground truth data for the detection performance. The frontal radar sensors and the camera have been measured during the supervised Bachelor thesis of Daniel Penning [221] and parts of the measurement results have been published together with first tracking results at a scientific conference [207].

The measurement space was fragmented into sensor-dependent grids with at least 100 measurements per grid cell (from laser scanner and measured sensor). Detections within a time frame around each reference measurement have been used for spatial association of sensor detections and ground truth data. In a first step, the association is performed using Cartesian vehicle coordinates of each measurement and the Euclidean distance as measure for association. Then, one computes preliminary covariance matrices dependent on detections in the measurement space based on the retrieved associations. In the following step, the association of ground truth data with the detections in the corresponding time frame is repeated in the measurement space using the Mahalanobis distance as measure for association. The final measurement covariance matrices $\mathbf{R}^S$ of the sensors are computed based on these associations for each grid cell:

$$\mathbf{z}_i^{\text{dev}} = \mathbf{z}_i^S - \mathbf{z}_i^{\text{ref}}, \tag{4.10}$$

$$\mu_{\mathbf{z}}^{\text{dev}} = \frac{1}{N_{\text{cell}}} \sum_{i=1}^{N_{\text{cell}}} \mathbf{z}_i^{\text{dev}}, \tag{4.11}$$

$$\mathbf{R}_{lm}^S = \frac{1}{N_{\text{cell}} - 1} \sum_{i=1}^{N_{\text{cell}}} (\mathbf{z}_{il}^{\text{dev}} - \mu_{\mathbf{z}}^{\text{dev}})(\mathbf{z}_{im}^{\text{dev}} - \mu_{\mathbf{z}}^{\text{dev}}), \tag{4.12}$$

where N_{cell} is the number of associations in the cell and the indexes l and m represent the corresponding rows of the measurement vector. $\mathbf{z}_i^S$ denotes the measurement vector from radar or camera and $\mathbf{z}_i^{\text{ref}}$ the reference measurement vector (ground truth).

The probability of detection $p_{\text{D}}(x_i)$ without consideration of occlusion and the false alarm probability $p_{\text{FP}}(z_j)$ have been determined under almost ideal conditions. Ratios have been computed using the reference measurements and associated detections of the corresponding grid cell according to:

$$p_{\text{D}}(x) = \frac{N_{\text{asso}}}{N_{\text{ref}}}, \tag{4.13}$$

$$p_{\text{FP}}(z) = 1 - \frac{N_{\text{asso}}}{N_{\text{det}}}, \tag{4.14}$$

where N_{asso} represents the number of detections that have been associated with a reference measurement, N_{ref} the number of reference measurements and N_{det} the number of detections from the measured sensor.

4.4.1 Measurement Results of the Camera

It has been mentioned in Section 4.2.4 that the dynamic pitch angle of the camera leads to some additional noise which cannot be neglected. The dynamic pitch angle due to acceleration and deceleration can be estimated and compensated, but the more high-frequent part of the pitch angle deviation, e.g., due to rough ground or gully covers, is taken into account as noise. In case of a pitch angle variance of $(1°)^2$, the variance in the vertical image coordinates v has to be increased by $(21\text{ px})^2$. The coordinate $[u, v] = [0, 0]$ would correspond to a point located far away on the left.

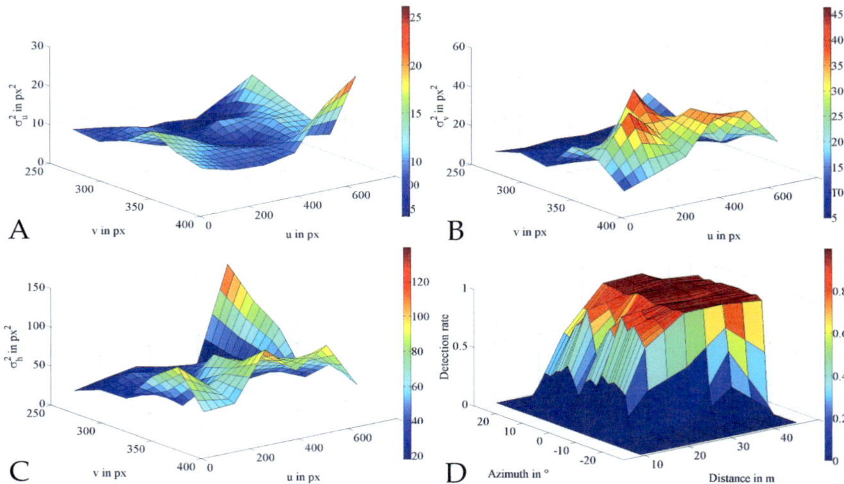

Figure 4.12   Variances and detection rate of adult pedestrians by the camera for a stationary ego vehicle.

The measurements have been performed with adult pedestrians and a stationary ego vehicle to evaluate the sensors' performance without dynamic impact. The variance σ_u^2 of the horizontal image coordinates u is $< 15\text{ px}^2$, see Figure 4.12. The higher inaccuracies in the lower right corner can result from lens distortion effects that could not be eliminated by calibration. The variance σ_v^2 of the vertical image coordinates v is $< 5\text{ px}^2$ in the upper part of the image and $< 35\text{ px}^2$ in the lower part of the image. The increased values in the near range result from the fact that pedestrian detections include more pixels in the near range and thereby a higher variation in pixels. The variance distribution of the object height σ_h^2 is similar to the variance of the vertical image coordinates in the way that it increases from about 20 px^2 in the far range to up to 60 px^2 in the near range. The inaccuracies are enlarged in the upper right part of the image. Although the camera controls the exposure time dependent on the illumination, the noise slightly increases during sunset, sunrise and illumination conditions with bad contrast. Under good measurement conditions, the detection rate within the FOV of the camera (10 m $< x <$ 45 m) is about constant ($p_{D,C} \approx 0.95$), see Figure 4.12 D. It decreases with worse illumination conditions, e.g., due to the motion blur, and in more crowded scenes with changing background. If there is a high number of pedestrians on the street, not all pedestrians are detected due to computational reasons. For example, if a large group of pedestrians crosses the street, the camera module will not detect all group members. The detection rate decreases to about 50% then. The detection rate decreases in the near range less steeply than on the side since some pedestrians are still detected although their foot point already disappeared.

4.4.2 Measurement Results of the LRR and SRR

The pitch angle deviation caused by rough ground or acceleration of the ego vehicle is not explicitly modeled for the radar sensors, since their distance measurement principle is not based on this angle. However, high pitch angles may reduce the detection performance, since the road surface then reflects the biggest part of the waves which is detected by the sensor. Therefore, the sensor does not provide any target measurements in these situations. The measurement variances have been determined using measurements from pedestrians that usually do not leave the grid cell during one measurement sequence. Thus, standing but not completely frozen pedestrians have been measured. Since the radar sensor is mounted only 0.39 cm above the ground and does not have a large vertical aperture angle, only pedestrian legs are detected, especially, in the near range. Thus, the angular variance and radial variance are higher for walking pedestrians. One defines a value for the standard deviation of pedestrian knees from the body center and takes this value into account by adding the resulting covariances to the ones obtained from standing pedestrians. The variance of the range rate is set constant and has been taken from the specification data sheet [11], since there are no reference measurements available for this signal. One has to adapt the variance of the range rate as well, since the measurements result from pedestrian legs that are either

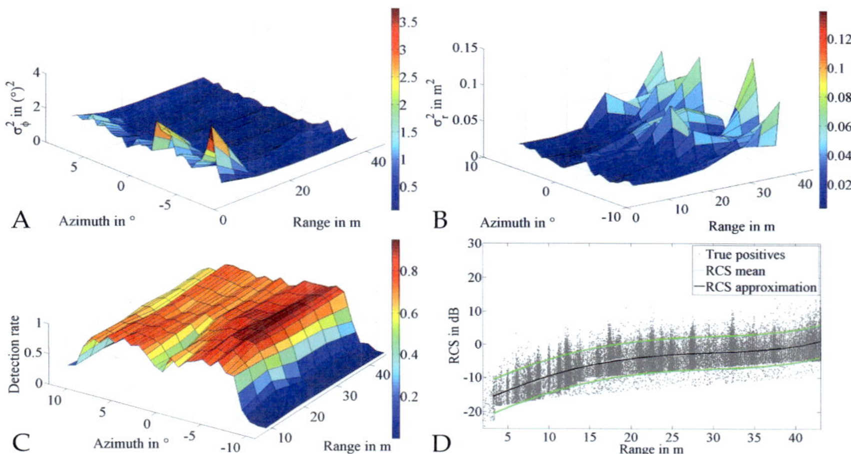

Figure 4.13 Determined angular and radial variance and detection rate (C) of the LRR for stationary moving pedestrians: The sensor detects only legs, especially in the near range. Additional variance in the pedestrian position due to walking is not included here, but added later. Subgraph D shows range-dependent RCS values of pedestrian measurements.

stationary or show values around the double pedestrian speed or more. Therefore, the variance values are chosen considerably higher then provided by the data sheet.[1]

The measurement results refer to results obtained from measurements with mainly standing pedestrians and do not include this additional variance. The variance of the azimuthal angle σ_Φ^2 decreases with the range and is not constant in the near range. It shows local maximums which can be explained by the arrangement and the characteristics of the antennas, see Figure 4.13 A. The range variance of the LRR is usually below 0.1 m^2 and is even lower in the near range, see Figure 4.13 B. The range accuracy of the SRR is comparable to the LRR within the same range except for the near range at the right side, where it increases a little bit.

The detection rate of the LRR $p_{\mathrm{D,LRR}}$ mounted at the front bumper is quite constant in wide regions (≈ 0.7) but decreases (down to 0.4) around $0°$ within the range up to 30 m, see Figure 4.13 C. A similar behavior can

---

[1]Suppliers usually measure their radar sensors using corner reflectors.

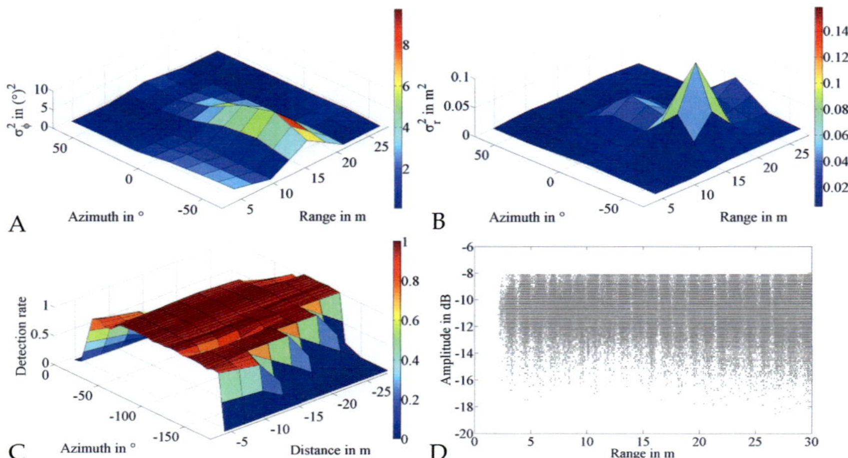

Figure 4.14 Determined angular and radial variance and detection rate (C) of the BSR for standing pedestrians: The sensor detects only legs, especially in the near range. Additional variance in the pedestrian position due to walking is not included here, but added later. Subgraph D shows the amplitude values of pedestrians detections over the range.

be observed for the SRR that is combined with the LRR in one housing. The SRR has a higher aperture angle but its angular accuracy is lower. The appendix contains plots of the measurement results of the SRR for standing pedestrians.

As mentioned above, there is a dependency between the false alarm rate of the radar and the obtained radar cross section (RCS) for pedestrians. A radar shows different interference waves in different heights. If an object approaches a sensor, it intersects different interference levels leading to a high variation in the RCS [135]. This behavior has been utilized to enable a kind of classification of the radar objects. Figure 4.13 D shows how the RCS values of the true positive measurements corresponding to pedestrians increase with the distance up to about 0 dB $\sim$ 1 m^2. The RCS values are smaller in the near range, since only legs can reflect within short distances.

4.4.3 Measurement Results of the BSR

The BSR has a higher aperture angle than the LRR and the angular accuracy of the single detections is consequently smaller but comparable to the SRR with less than $3(^\circ)^2$. The variance of the range accuracy is within a few centimeters, see Figure 4.14 A and B. The sensor has been designed for the detection of vulnerable road users, but it does not detect the whole body in the near range due its low mounting position. In contrast to the frontal radar sensors, the BSR usually detects both legs, so that one can determine the body position between the legs. There are often up to six detections from one pedestrian due to different speeds of the pedestrian's extremities (separation by Doppler speed). Two detections from one pedestrian are the minimum, whereas other objects evoke only single detections. Thus, the number of detections from one pedestrian and the detection distances to each other can be applied as classification attribute for pedestrians. Then, the detections are grouped to one target. The mean values of the single detection properties are taken as measurement properties of the resulting target. The detection rate of pedestrians by the BSR is about constant and very high — more than 90 % in a wide range, see Figure 4.14 C.

The amplitude value was intended to serve as BSR feature for pedestrian classification — comparable to the RCS value of the frontal radar sensors. It should represent the reflectivity of a detected target. Thus, reflections from cars should show higher values than reflections from pedestrians. The amplitudes provided by the sensor of various target types were all located within the same range (see Figure 4.14 D), so that this feature is not very useful and is not considered any further.

4.5 Process and Measurement Models

Since vehicles are limited in their rotational capabilities due to their architecture, a state model with constant turn rate and constant velocity (CTRV) based on the single track model is applied for vehicles — including the ego vehicle. In case of very small yaw rate values, this state model is reduced to a constant velocity and constant orientation model (CVCO) due to singularity. The origin of the local vehicle coordinate system at time point k is located in the middle of the back axle on the ground.

A constant speed model is chosen for pedestrians due to their increased rotational degree of freedom. Beside the position vector $[x,\ y]^{\mathrm{T}}$, one is interested in the pedestrian's longitudinal and lateral speed $[v_x,\ v_y]^{\mathrm{T}}$ over ground in the world as well as in his height s (size). The width is not tracked, since the shapes of pedestrians change too much during motion (walking, running), making it hard to define the pedestrian width then. Therefore, the pedestrians are modeled as vertical line objects in the process model, although their dimension is taken into account in the birth model of the JIPDA filter. The pedestrians are assumed to follow a linear state model with constant longitudinal and lateral speed. The object state vector at time point k corresponds to $\mathbf{x}_k = [x,\ y,\ v_x,\ v_y,\ s]^\mathrm{T}$.

4.5.1 Pedestrian State Model

The presented tracking approach serves the state estimation of multiple pedestrians in the environment. The prior state estimation $\hat{\mathbf{x}}_{k|k-1}$ in local vehicle coordinates for a stationary ego vehicle at time point k can be computed using the constant speed assumption, so that

$$\hat{\mathbf{x}}_{k|k-1} = f_{\mathrm{ped}}\left(\hat{\mathbf{x}}_{k-1|k-1}\right) \tag{4.15}$$

$$= \begin{bmatrix} x_{k|k-1} \\ y_{k|k-1} \\ v_{x,k|k-1} \\ v_{y,k|k-1} \\ s_{k|k-1} \end{bmatrix} = \begin{bmatrix} x_{k-1|k-1} + v_{x,k-1|k-1} \cdot \Delta t \\ y_{k-1|k-1} + v_{y,k-1|k-1} \cdot \Delta t \\ v_{x,k-1|k-1} \\ v_{y,k-1|k-1} \\ s_{k-1|k-1} \end{bmatrix}, \tag{4.16}$$

where Δt is the time increment representing the time between the last update and the current measurement. The vector $[x,\ y]^\mathrm{T}$ represents the pedestrian's position in Cartesian vehicle coordinates, v_x and v_y her speed components in the world and s her height (size). As long as no ego motion is taken into account, the process model is linear and no Jacobian matrix has to be computed.

Since the process model neglects all terms of higher order such as the pedestrian acceleration, the process model error is included in the normally distributed process noise $\omega_{\mathrm{ped}} \sim \mathcal{N}\left(0^5, \mathbf{Q}_{\mathrm{ped}}\right)$ with covariance matrix

$$\mathbf{Q}_{\mathrm{ped}} = \begin{bmatrix} \frac{1}{3}\Delta t^2 \sigma_{v_x}^2 & 0 & \frac{1}{2}\Delta t \sigma_{v_x}^2 & 0 & 0 \\ 0 & \frac{1}{3}\Delta t^2 \sigma_{v_y}^2 & 0 & \frac{1}{2}\Delta t \sigma_{v_y}^2 & 0 \\ \frac{1}{2}\Delta t \sigma_{v_x}^2 & 0 & \sigma_{v_x}^2 & 0 & 0 \\ 0 & \frac{1}{2}\Delta t \sigma_{v_y}^2 & 0 & \sigma_{v_y}^2 & 0 \\ 0 & 0 & 0 & 0 & \sigma_s^2 \end{bmatrix}. \tag{4.17}$$

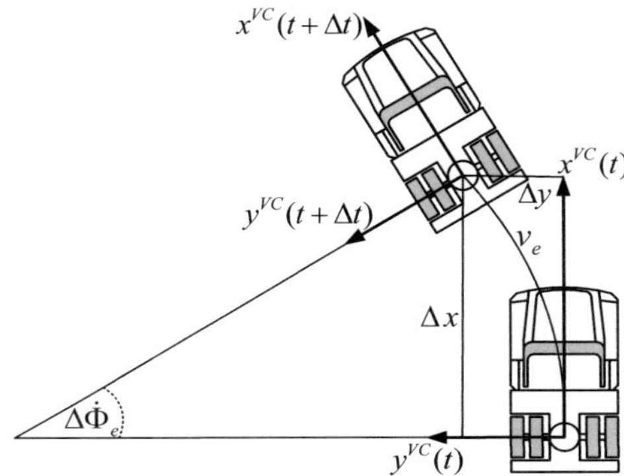

Figure 4.15 Single Track Model.

4.5.2 Ego Motion Estimation

State variables of other road users can only be described relatively to the coordinate system of the stationary ego vehicle so far. However, the ego vehicle is usually not stationary but moves between different measurement cycles. This has to be considered in the prior state estimation and in the process model.

Wheel slip and drift are neglected and extreme driving maneuvers are not considered here. The time-dependent tangential speed $v_{e,k}$ and the yaw rate around the vertical axis $\dot{\phi}_{e,k}$ describe the ego vehicle's dynamics based on the single track model, see Figure 4.15. The state model for vehicles is based on the assumption of constant speed and constant turn. The additional input vector $\mathbf{u}_{e,k}$ compensates for the ego vehicle's translation between the two time steps $k-1$ and k:

$$
\mathbf{u}_{e,k} = \frac{v_{e,k-1}}{\dot{\phi}_{e,k-1}}
\begin{bmatrix}
\sin\left(\dot{\phi}_{e,k-1}\Delta t\right) \\
1 - \cos\left(\dot{\phi}_{e,k-1}\Delta t\right) \\
0 \\
0 \\
0
\end{bmatrix} .
\tag{4.18}
$$

The rotation matrix $\mathbf{R}_{e,k}$ takes into account the vehicle's rotation around its vertical axis within the corresponding time interval:

$$
\mathbf{R}_{e,k} =
\begin{bmatrix}
\cos\left(\dot{\phi}_{e,k-1}\Delta t\right) & -\sin\left(\dot{\phi}_{e,k-1}\Delta t\right) \\
\sin\left(\dot{\phi}_{e,k-1}\Delta t\right) & \cos\left(\dot{\phi}_{e,k-1}\Delta t\right)
\end{bmatrix} .
\tag{4.19}
$$

Derivatives of higher order like the acceleration are neglected, which is considered in the normally distributed process noise $w_e \sim \mathcal{N}(0^2, \mathbf{Q}_e)$ with

$$\mathbf{Q}_e = \begin{bmatrix} \sigma_{v_e}^2 & \sigma_{v_e}\sigma_{\dot{\phi}_e} \\ \sigma_{\dot{\phi}_e}\sigma_{v_e} & \sigma_{\dot{\phi}_e}^2 \end{bmatrix}. \tag{4.20}$$

Consequently, the prior state estimation $\hat{\mathbf{x}}_{k|k-1}$ for time step k can be obtained using the discrete process model $f(\hat{\mathbf{x}}_{k-1|k-1})$ for the pedestrian and the ego vehicle:

$$\hat{\mathbf{x}}_{k|k-1} = f(\hat{\mathbf{x}}_{k-1|k-1}) \tag{4.21}$$

$$= \begin{bmatrix} \cos(\Delta\phi_e)x + \sin(\Delta\phi_e)y - v_e(\dot{\phi}_e)^{-1}\sin(\Delta\phi_e) \\ -\sin(\Delta\phi_e)x + \cos(\Delta\phi_e)y - v_e(\dot{\phi}_e)^{-1}[\cos(\Delta\phi_e) - 1] \\ \cos(\Delta\phi_e)v_x + \sin(\Delta\phi_e)v_y \\ -\sin(\Delta\phi_e)v_x + \cos(\Delta\phi_e)v_y \\ s \end{bmatrix},$$

$$\tag{4.22}$$

where the time indexes k have been neglected and $\Delta\phi_e$ corresponds to $\dot{\phi}_e\Delta t$.

The Jacobian matrix $\mathbf{F}_k$ of the process model

$$\mathbf{F}_k = \left.\frac{\partial f}{\partial(x, y, v_x, v_y, s, v_e, \dot{\phi}_e)}\right|_{\hat{\mathbf{x}}_{k-1|k-1}, \hat{v}_e, \hat{\dot{\phi}}_e} \tag{4.23}$$

serves the computation of the prior estimation of the covariance matrix $\hat{\mathbf{P}}_{k|k-1}$ of the state:

$$\hat{\mathbf{P}}_{k|k-1} = \mathbf{F}_k \begin{bmatrix} \mathbf{Q}_{\text{ped}} & 0 \\ 0 & \mathbf{Q}_e \end{bmatrix} \mathbf{F}_k^{\text{T}}, \tag{4.24}$$

where the uncertainty of the state is propagated.

The special case of an evanescent yaw rate of the ego vehicle (straight driving) leads to the fact that the first two rows of $f(\hat{\mathbf{x}}_{k-1|k-1})$ are undetermined. Therefore, the limit function $f_0(\hat{\mathbf{x}}_{k|k-1})$ and its Jacobian are applied instead if the absolute value of the yaw rate $\dot{\phi}_e$ is below a threshold value $\epsilon_{\dot{\phi}_e}$:

$$\hat{\mathbf{x}}_{k|k-1} = f_0(\hat{\mathbf{x}}_{k|k-1}) \tag{4.25}$$

$$= \lim_{\dot{\phi}_e \to 0} f(\hat{\mathbf{x}}_{k|k-1}) = [x - v_e\Delta t, \ y, \ v_x, \ v_y, \ s]^{\text{T}}. \tag{4.26}$$

The pitch behavior of the ego vehicle due to rough ground or acceleration impacts the estimation of the object's longitudinal coordinate x. It can be modeled as rotation around the y axis of the vehicle's coordinate system. However, there are no sensors available here that measure the vehicle's pitch angle. In this work, the dynamic pitch angle component that is induced by speed and acceleration of the ego vehicle is tried to be compensated for by assuming a linear relation between the pitch angle and the acceleration. All other effects that lead to a dynamic pitch angle are sub-summarized in the measurement noise of the sensors radar and camera.

4.5.3 Radar Measurement Model

The predicted state needs to be converted to the measurement space for data association. Therefore, the prior state estimation $\hat{x}_{k|k-1}$ in Cartesian vehicle coordinates of time point k is transformed to the sensor coordinate system using the transformation matrix $\mathbf{T}_{\text{SC2VC}}^{-1}$ from Equation 4.3. $\hat{x}_{k|k-1}^{\text{S}}$ denotes the prior state estimation in Cartesian sensor coordinates.

In case of the radar, the transformation matrix contains the translation vector $\mathbf{t}^{\text{R}}$ and the rotation matrix $\mathbf{R}_{\text{rot}}^{\text{R}}$. Each radar measurement provides the range r to a detected target in m, the bearing angle ϕ in $°$ and the range rate $\dot{r}$ in m/s. Since the range rate is a relative velocity measure, the components $v_{x,k|k-1}$ and $v_{y,k|k-1}$ representing the speed over ground have to be transformed to relative speed components $v_{x,k|k-1}^{\text{R}}$ and $v_{y,k|k-1}^{\text{R}}$ depending on the ego vehicle's motion, so that

$$\hat{x}_{k|k-1}^{\text{R}} = \begin{bmatrix} x_{k|k-1}^{\text{R}} \\ y_{k|k-1}^{\text{R}} \\ v_{x,k|k-1}^{\text{R}} \\ v_{y,k|k-1}^{\text{R}} \\ s_{k|k-1}^{\text{R}} \end{bmatrix} = \begin{bmatrix} r_{11} \cdot x_{k|k-1} + r_{12} \cdot y_{k|k-1} + t_x \\ r_{21} \cdot x_{k|k-1} + r_{22} \cdot y_{k|k-1} + t_y \\ r_{11} \cdot (v_{x,k|k-1} - v_{e,x}) + r_{12} \cdot (v_{y,k|k-1} - v_{e,y}) \\ r_{21} \cdot (v_{x,k|k-1} - v_{e,x}) + r_{22} \cdot (v_{y,k|k-1} - v_{e,y}) \\ s_{k|k-1} \end{bmatrix}$$

$$(4.27)$$

where r_{lm} and $t_{\square}$ represent the entries of the rotation matrix and the translation vector, respectively. $v_{e,x}$ and $v_{e,y}$ denote the longitudinal and lateral speed of the ego vehicle, where the time indexes have been neglected. Then, the non-linear radar measurement function $h^{\text{R}}(\hat{x}_{k|k-1}^{\text{R}})$ transforms the prior state estimation to the measurement space according to

$$h^{\text{R}}(\hat{x}^{\text{R}}) = [r(\hat{x}^{\text{R}}),\ \phi(\hat{x}^{\text{R}}),\ \dot{r}(\hat{x}^{\text{R}})]^{\text{T}}, \qquad (4.28)$$

where

$$
r(\hat{\mathbf{x}}^{\mathrm{R}}) \;=\; \sqrt{(x^{\mathrm{R}})^2 + (y^{\mathrm{R}})^2}, \tag{4.29}
$$

$$
\phi(\hat{\mathbf{x}}^{\mathrm{R}}) \;=\;
\begin{cases}
\arctan\left(\frac{y^{\mathrm{R}}}{x^{\mathrm{R}}}\right), & \text{if } x^{\mathrm{R}} \geq 0, \\[2mm]
\arctan\left(\frac{y^{\mathrm{R}}}{x^{\mathrm{R}}}\right) + 180^\circ, & \text{if } x^{\mathrm{R}} < 0 \wedge y^{\mathrm{R}} \geq 0, \\[2mm]
\arctan\left(\frac{y^{\mathrm{R}}}{x^{\mathrm{R}}}\right) - 180^\circ, & \text{if } x^{\mathrm{R}} < 0 \wedge y^{\mathrm{R}} < 0,
\end{cases} \tag{4.30}
$$

$$
\dot{r}(\hat{\mathbf{x}}^{\mathrm{R}}) \;=\; \sqrt{(v_x^{\mathrm{R}})^2 + (v_y^{\mathrm{R}})^2} \cdot \sin\left(\phi(\hat{\mathbf{x}}^{\mathrm{R}}) + \phi_v(\hat{\mathbf{x}}^{\mathrm{R}})\right), \tag{4.31}
$$

where

$$
\phi_v(\hat{\mathbf{x}}^{\mathrm{R}}) =
\begin{cases}
\arctan\left(\frac{v_y^{\mathrm{R}}}{v_x^{\mathrm{R}}}\right), & \text{if } v_x^{\mathrm{R}} \geq 0, \\[3mm]
\arctan\left(\frac{v_y^{\mathrm{R}}}{v_x^{\mathrm{R}}}\right) + 180^\circ, & \text{if } v_x^{\mathrm{R}} < 0 \wedge v_y^{\mathrm{R}} \geq 0, \\[3mm]
\arctan\left(\frac{v_y^{\mathrm{R}}}{v_x^{\mathrm{R}}}\right) - 180^\circ, & \text{if } v_x^{\mathrm{R}} < 0 \wedge v_y^{\mathrm{R}} < 0.
\end{cases} \tag{4.32}
$$

Since the angle for tracking is larger than 180° (π), one has to resolve ambiguities in the angle computation. The measurement function is non-linear due to the quadratic terms, the fractions, and the trigonometric components. However, the application of the EKF requires a linear function, so that the non-linear measurement function $h^{\mathrm{R}}(\hat{\mathbf{x}}_{k|k-1}^{\mathrm{R}})$ is linearized at $\hat{\mathbf{x}}_{k|k-1}^{\mathrm{R}}$ using the Jacobian matrix $\mathbf{H}_k^{\mathrm{R}}$:

$$
\mathbf{H}_k^{\mathrm{R}} = \left. \frac{\partial h^{\mathrm{R}}(\mathbf{x})}{\partial(x, y, v_x, v_y, s)} \right|_{\mathbf{x} = \hat{\mathbf{x}}_{k|k-1}^{\mathrm{R}}}. \tag{4.33}
$$

4.5.4 Camera Measurement Model

The HOG classifier running on the camera image provides the pedestrian hypotheses (measurements) in image coordinates (pixels). The origin of this coordinate system is located in the upper left corner of the image. The object position in the image is represented by the coordinate vector $[u, v]^{\mathrm{T}}$ which is the foot point of the pedestrian in the middle of the detected box (half box width), and the pedestrian height s^{I} in pixels. The pedestrian is

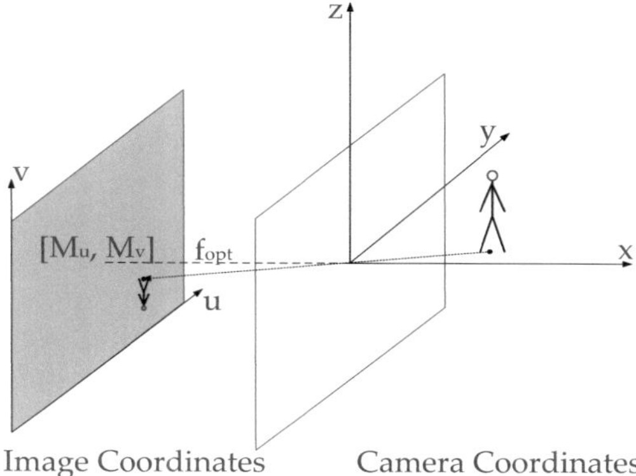

Figure 4.16 Camera pinhole model: The three-dimensional camera coordinates are projected onto the two-dimensional image plane where the intercept theorem holds.

expected to stand on the ground plane. The association in the measurement space requires the transform of the prior state estimation $\hat{\mathbf{x}}_{k|k-1}$ from the current ego vehicle coordinate system to the Cartesian camera coordinate system ($\hat{\mathbf{x}}^C_{k|k-1}$) using the transformation matrix $\mathbf{T}^{-1}_{\text{CC2VC}}$. Next, one has to project the state estimation $\hat{\mathbf{x}}^C_{k|k-1}$ to image coordinates. Since the lens distortion is corrected using the parameters of the intrinsic calibration before any computations are performed on the image frames (e.g., HOG detections), the measurement function can be obtained from the pinhole camera model, see Figure 4.16. The pinhole model requires the values of the optical center $[M_u, M_v]^T$ and the focal length f_{opt} as parameters of the measurement function $h^C(\hat{\mathbf{x}}^C_{k|k-1})$ that projects the prior state estimation to image coordinates:

$$h^C(\hat{\mathbf{x}}^C) = [u(\hat{\mathbf{x}}^C),\, v(\hat{\mathbf{x}}^C),\, h(\hat{\mathbf{x}}^C)]^T, \tag{4.34}$$

where

$$u(\hat{\mathbf{x}}^C) = -f_{\text{opt}}\frac{y^C}{x^C} + M_u, \tag{4.35}$$

$$v(\hat{\mathbf{x}}^C) = -f_{\text{opt}}\frac{z^C}{x^C} + M_v, \tag{4.36}$$

$$s^I(\hat{\mathbf{x}}^C) = f_{\text{opt}}\frac{s^C}{x^C}. \tag{4.37}$$

As for the radar, the measurement function of the camera is non-linear, so that it has to be linearized at $\hat{\mathbf{x}}^C_{k|k-1}$ for application in the EKF using the Jacobian $\mathbf{H}^C_k$ of the measurement function $h^C(\hat{\mathbf{x}}^C_{k|k-1})$:

$$\mathbf{H}^C_k = \left. \frac{\partial h^C(\mathbf{x})}{\partial(x, y, v_x, v_y, s)} \right|_{\mathbf{x}=\hat{\mathbf{x}}^C_{k|k-1}} . \tag{4.38}$$

The inverse transform of the measurement function, requires the application of the homography condition. It assumes that the pedestrians are located on the ground and that the ground is flat ($z = 0$). Furthermore, the correctness of the inverse transformation strongly depends on the correct estimation of the rotation angles, especially the pitch angle.

4.5.5 Existence Models

All sensors underlie uncertainties and false positive detections can never be excluded. Nonetheless, one desires to maximize the detection rate of a sensor system, but its false positive rate shall be minimized. If active applications should react to an object based on the data from the environment perception module, the confidence that the object really exists has to be high. Standard approaches based on LNN or GNN usually consider only measurements with features exceeding some threshold, e.g., the HOG confidence value of the camera. An object would only be handed over to the application if measurements have been assigned to the object for a certain number of cycles. Sometimes a heuristic confidence value of the object is provided based on the measurement cycles since the time of object birth and the corresponding number of associated measurements. This approach is straightforward, but the detection rate due to previous thresholding decreases with a reduced false positive rate. When one track is lost and detected again, a new ID is assigned, so that a single physical object might get assigned numerous IDs during its observation. This is often acceptable if the algorithm output is only used for one application and only a short data history is required in the situation evaluation module. In case that several applications process their data and constant IDs for a physical object are desired, it would be more helpful to track the objects based on all available measurements and provide a PoE for each object, so that the subsequent applications can decide if the object confidence suffices for a reaction or not. This requires that the PoE be more than just a heuristic value and motivates the application of the

JIPDA, where the object existence is tracked based on a persistence model and an existence measurement model.

Existence Models using the JIPDA

The object persistence model corresponds to the process model for existence estimation in the JIPDA. The prior PoE of the object for time step k is predicted based on the PoE of the previous time step $k-1$ and the probability of survival $p_S(x_i)$ (compare Section 3.2.2). Objects that exceed state constraints, for instance such ones having an abnormally high speed ($|v| > 4\,\mathrm{m/s}$) or being located outside the FOV should have a low chance to survive. The probability of survival $p_S(x_i)$ is set to an experimentally obtained value of 98 % otherwise, so that

$$p_{k|k-1}(\exists x_i) = p_S(x_i) \cdot p_{k-1|k-1}(\exists x_i), \tag{4.39}$$

where

$$p_S(x_i) = \begin{cases} 0.00001, & \text{if } \sqrt{v_{x,k-1|k-1}^2 + v_{y,k-1|k-1}^2} > v_{\max}, \\ 0.00001, & \text{if } x_i \text{ outside tracking region,} \\ 0.98, & \text{else.} \end{cases} \tag{4.40}$$

The existence model quantifies existence information of a sensor based on object measurements. The probability of object existence is modeled based on the sensory inference probability $p_{TP}(z_j)$ that describes the false alarm process depending on measurement attributes as well as on the probability that there will be no measurement although the object exists $(1 - p_D(x_i))$. Furthermore, the probability of object birth is modeled and provides a valuable support for the track management.

Data-based statistic approaches are used to determine the sensor characteristic in this work. The laser scanner measurements serve as reference data, as described in Section 4.2.4. The sensory inference probability $p_{TP}(z_j)$ of the camera is determined based on the value that is provided by the ROC curve of the HOG classifier and subsequent NMS. The measurements of the front radars contribute in different way to the existence estimation, since the provided RCS values are a criteria for pedestrians but do not contain additional information. Only measurements with RCS values within the measured band are processed, but no sensory inference probability is computed based on this value. A constant value $p_{FP}(z_j) = 0.25$ is assumed for each measurement. An existence estimation $p_{k|k}^*(\exists x_i)$ is computed based on this value. If the resulting value exceeds the prior existence estimate $p_{k|k-1}(\exists x_i)$, it is assumed that there has been a detection of the object and the posterior estimate is set to the value of the prior estimate $p_{k|k}(\exists x_i) = p_{k|k-1}(\exists x_i)$. Otherwise, the posterior value is set to the computed lowered PoE $p_{k|k}(\exists x_i) = p^*_{k|k}(\exists x_i)$. Consequently, missing detections can lower the PoE of an object, but associated measurements only rarely impact the existence estimation. Thereby, the insufficient classification ability of the radar can be compensated.

Correct modeling of the state-dependent detection probability $p_D(x_i)$ is essential for the detection performance of the total system. Rough inaccuracies, e.g., due to an inaccurate calibration, evoke that tracked obstacles are discarded or that their PoE rises too slowly.

Usually, the detection probability p_D depends on the pedestrian state, e.g., on the pedestrian position in the FOV of the sensor. The estimation problem reveals that the object state x_i is unknown, so that one considers the detection probability as conditional PDF $p_D(x_i|\mathbf{x}_i)$ of the variate $\mathbf{x}_i \sim \mathcal{N}(\mathbf{x}_{i,k|k-1}, \hat{\mathbf{x}}_{i,k|k-1}, \hat{\mathbf{P}}_{i,k|k-1})$. However, the JIPDA approach expects the non-conditional PDF of the detection process, which can be obtained using the law of total probability [65]:

$$p_D(x_i) = \int_{-\infty}^{\infty} p_D(x_i|\hat{\mathbf{x}}_i) p_D(\hat{\mathbf{x}}_i) \mathrm{d}\hat{\mathbf{x}}_i \tag{4.41}$$

$$= \int_{-\infty}^{\infty} \ldots \int_{-\infty}^{\infty} p_D(x_i|x, y, \ldots) p_D(x, \ldots, s) \, \mathrm{d}x \ldots \mathrm{d}s. \tag{4.42}$$

To the author's experience, the detection capability of the sensors depend on the pedestrian positions for the camera and not on their speed or orientation. The size of a pedestrian is a determining factor as well, since the maximum detection range of taller pedestrians is higher. It is expected that the detection probability decreases (for instance, -10% for pedestrians shorter than 1.5 m compared to taller pedestrians). This assumption was not validated, as no children were available as test persons.

Objects can be separated more easily from other objects by the radar if they differ in speed. However, no speed dependency could be observed for free standing or moving pedestrians. Moreover, the reference sensor provides a ground truth for the object position but not for the speed, so that the detection probability $p_D(x_i)$ is only modeled dependent on the object position (refer to previous sections for illustration).

Different sensors like radar and the monocular camera consider different objects as relevant. The radar detects various object types but does not classify them. It does not provide a reliable measure for the sensory inference probability $p_{\mathrm{TP}}(z_j)$ of a pedestrian, while the camera system only detects the objects for which its classifier has been trained. Since one is only interested in the relevant objects (e.g., pedestrians), only these should be tracked and should have assigned a high PoE.[2] Here, the conflict is solved by reducing the impact of the front radars' detections on the existence update, whereas missing detections can lower the PoE of an object as described above.

4.6 Track Management

4.6.1 EKF-JIPDA Track Management

An important step in the tracking procedure is the track management. Unobserved objects, e.g., as they just entered the FOV, require initialization when they newly appear and objects that disappeared or receive no more measurements have to be removed from the track list.

The EKF with JIPDA implicitly includes the track management, as the probability of birth is estimated from measurement-based object hypotheses and the existing environment model. A new object is set up if there is no other object within the same gating region or the probability of birth exceeds the PoE that would result if the measurement was assigned to another object. Object removal from the track list exploits the object's PoE. If the PoE falls below a certain threshold value (< 0.0001), the object is deleted. Thus, no special heuristics are required.

The low reflection values of pedestrians and the limited classification ability of the radar lead to the fact that reflections from other weakly reflecting objects have to be accepted as well. Therefore, no application would accept a track that is only based on weakly reflecting targets from the frontal radars. This differs for tracks resulting from strongly reflecting obstacles like cars and thus, higher object confidences. Since a camera confirmation would be required for any reaction to pedestrians, only the camera is allowed to initialize a pedestrian track in the front area and the radar information is only used for an improved state estimation, track confirmation and to lower the PoE in case of missing detections. One

---

[2]Munz [65] separated the object existence into a relevant and an irrelevant part developing an extension of Dempster's and Shafer's theory of evidence. He took into account that not all sensors are able to distinguish between the two classes relevant and irrelevant, and that they are ignorant which information other sensors provide. The implementation of this approach for pedestrians could be content of future work.

advantage of this procedure is that the computational complexity is reduced because less objects have to be tracked. As the LRR and the SRR usually detect only one of the two legs and the accuracy of the leg speed is very low, especially, in case of lateral pedestrian motion, this information cannot be used for the initialization of the track and an initialization with speed 0 is favored. If the algorithm detects the front leg in the first measurement cycle and the retral leg in the next cycle, the track would be initialized with a speed component to the wrong direction if only two measurement cycles were used. Therefore, at least three measurements from the same pedestrian would be required to initialize a track if only one leg is detected at once.

In case of the BSR, at least both pedestrian legs are detected providing a higher pedestrian confidence and enabling a radar-based track initialization. Usually, there are more than three detections per pedestrian. As pedestrian speeds are usually low, the pedestrian tracks are initialized based on one measurement as stationary obstacles with a high uncertainty in the speed-dependent components of the state covariance matrix instead of applying a multiple-step initialization.

4.6.2 EKF-GNN Track Management and Object Confidence Computation

The EKF with JIPDA (EKF-JIPDA) provides the PoE of an object as confidence value which can be used as criterion for the track management. This is not the case for the EKF with GNN (EKF-GNN). The EKF-GNN only processes measurements that satisfy the HOG-based threshold or have their RCS value within the defined band. Then an object has to be assigned to measurements at least for a certain number of cycles ϵ_{N_a} in a row for confirmation. Another criterion for object confirmation is that the fraction of associations and total cycles has to exceed a certain threshold value. An object is deleted from the object list if there has not been any association for more than $\epsilon_{N_{na}}$ measurement cycles or the fraction of missing associations becomes too high. These values have to be set dependent on the object's position, since there have to be expected less associations in the sensory blind region than within the fields of detection. The data association for the EKF-GNN and the corresponding track management have been implemented during the supervised Diploma thesis of Wladimir Gerber [222].

The object confidence that is often desired in subsequent situation evaluation modules is determined as ratio of the number of measurements associated with the object since its birth N_a and the number of cycles since object birth $N_a + N_{na}$ (object age). This confidence value is often sufficient for objects with a low age but loses its meaning for longer tracks. It does neither take into account the object state nor the measurement properties. If a track should be kept alive across a sensory blind region, a heuristic has to define how long a track may stay alive without measurement association in this region, e.g., dependent on the speed of the ego vehicle.

Apart from that, initialization is comparable to the EKF-JIPDA. Objects in the vehicle front may only be initialized by the camera, while the BSR is allowed to initialize tracks within its FOV.

4.7 Tracking across the Sensory Blind Region

Figure 4.1 B visualizes that there is no common FOV of the LRR, the SRR and the camera with the BSR but a sensory blind region between the frontal FOVs and the FOV on the side. It would be advantageous for some applications to obtain some information about that region as well, but there are even more reasons why it is worth to track across the blind regions instead of using separate filters for the front and for the side. First, the camera-based classification is a big advantage for pedestrian detection, since radars can only barely distinguish between pedestrians and other weakly reflecting objects. If the objects from the front were handed over to the side, this classification knowledge could be conserved. Moreover, it always takes some time until a track is confirmed and the certainty of the state estimation has increased. If only one track has to be set up, the confirmation time could be shortened and earlier reactions would be possible in critical situations. Moreover, situation evaluation modules prefer one constant ID for one physical object (instead of two), which is not possible for separate tracking filters. Therefore, tracking across the blind region is desired.

If there are no more measurement associations, the values of the state covariance matrix grow when using an EKF, while the state is predicted and estimated based on the process model. Potential accelerations and a potential direction change are taken into account by the process noise propagation.

Since no sensor can detect an object in the sensory blind region, the probability of detection p_D is zero in this region. In case of the JIPDA-EKF, the PoE of objects in this region will decrease as well. The node probabilities $p(x_i, z_j) = 0$ and $p(x_i, \varnothing) = p_{k|k-1}(\exists x_i)$ lead to the fact that the update step is not able to provide additional information and expects the object still to exist, while the existence prediction lowers the PoE by multiplication with the persistence probability p_S. If the object's PoE falls below a certain threshold level, the object is deleted from the environment model. Thus, if an object with a high PoE enters the sensory blind region, it will survive longer than an object with a lower PoE.

When the object enters the FOV of the next sensor, it can be updated again based on the measurement data. Hence, the tracking probability of an object across the blind region depends on the conformity between the real object motion and the (linear) prediction model, on the object's PoE when it left the former FOV, the provided probability to survive p_S and the measurement properties when it reappears in the new FOV. A generic implementation is realized by virtually enlarging the FOV of the BSR in such way that it covers the blind region.

Occlusion can be modeled in a similar way. The region of occlusion with a small probability of detection can be represented by a Gaussian distribution that depends on the covariance matrix and position (and dimensions) of the occluding object. Furthermore, the detection probability of a pedestrian by the radar could be reduced if the pedestrian approaches an object that provides a reflection with a high RCS value that covers the pedestrian echo, e.g., a car. If the camera detects a pedestrian and the radar shows a high RCS value at the same position, one can reduce the radars' probability of detection and increase it with increasing distance of the pedestrian from the car [207].

5 Situation Evaluation

Chapter 4 described the creation of an environment model containing pedestrians. The objects' state estimations and the corresponding uncertainties have been determined as well as the objects' PoEs. However, the mere knowledge about pedestrian states in the environment does not help to avoid collisions. A situation evaluation module has to assess the criticality of the objects (pedestrians) and the situation. A favorable approach is to predict the future states of the ego vehicle including the driver behavior as well as the future states of the pedestrians. Next, the possible whereabouts of both road users at equal time points are intersected to assess the collision risk. The whereabouts can be based on deterministic positions or follow a type of probability distribution, so that the intersection has to take this into account. A similar approach is applied here. Probable paths of the road users are computed for a temporal prediction horizon of up to three seconds in a first step. Then the future states of the road users along these paths are computed for time points and time intervals using Monte-Carlo techniques and Markov chain abstraction, which leads to reachable sets. Finally, a conditional collision probability is computed by intersection of these reachable sets. The approach takes into account the path probabilities, potential deviation from the corresponding paths and uncertainty regarding the dynamics along the corresponding path. Furthermore, the PoEs of tracked objects as well as the variance in the state estimation are incorporated.

Many accidents could be avoided or reduced in severity if the driver is warned three seconds before a potential collision. Furthermore, non-avoidable false alarms, inaccuracies in the environment perception, and uncertainty in the situation prediction as well as the requirement that the driver must keep the control of the system at all times demands a warning phase — optical, acoustical and/or haptical — before an active system reaction. The driver must have the chance to overrule the system, to brake on his own, to initiate an evasion maneuver or to cancel the system reaction to avoid a collision with an obstacle. Moreover, the system warns the following traffic by activating the warning lights to avoid rear-end collisions with the following traffic. The system should initiate a warning phase about three seconds before a potential collision due to these requirements and based on results from accident analysis.

The next section starts with a summary of the state of the art in the field of related situation assessment and explains the deficits of the existing approaches. The first subsection is dedicated to approaches for maneuver classification and trajectory prediction, see Subsection 5.1.1, whereas the expression *maneuver classification* is also referred to as *driver intention prediction* in literature. The subsequent Subsection 5.1.2 gives an overview of existing methods that have been proposed for risk assessment in the last decade. Section 5.2 introduces a novel approach to predict the ego vehicle's path based on previous maneuver classification, while Section 5.3 describes the proposal for the computation of reachable sets and their intersection for the single road users. The resulting conditional crash probability is applicable as a new criticality measure in combination with a threshold to decide about the initiation of an active system reaction.

5.1 State of the Art regarding Situation Assessment

5.1.1 Related Work on Maneuver Classification and Trajectory Prediction

Standard approaches for motion prediction apply standard filter methods like the Kalman filter and forecast the vehicle's position based on a recursive prediction of the system state to the next time step until the desired prediction horizon is reached. Many common model-based methods assume constant velocities and constant yaw rates, which can result in large deviations from the real trajectory, especially in case of turning maneuvers.

Long-term behavior prediction and classification has been intensively studied in surveillance context [136–140], i.e., for anomaly detection. Those approaches usually use object motion patterns reflecting the knowledge of the scene and information about the fixed environment. Since vehicles move within the environment, so that background and motion patterns change, these requirements are not met anymore.

Approaches based on a long-term motion prediction for urban crossing scenarios with moving systems have been presented in [141, 142]. There, the motion patterns of vehicles are represented by trajectories defined as ordered tuples. The approach learns motion patterns by building a motion database consisting of observed trajectories and measures the similarity between trajectories. A particle filter framework generates a large number of motion hypotheses and assigns a likelihood value to them.

The drawback is that the environment has to be known exactly and many trajectories have to be stored in large databases and accessed online. If the road geometry differs from the one stored in the database, the system might not perform well. Keller [143] used a similar method to predict a probabilistic pedestrian path based on learned motion features that are obtained from a stereo camera. The potential of his approach to judge whether or not a pedestrian will cross the lane increased the one of other approaches, but the method was not able to predict the behavior correctly in one half of the situations.

Dagli et al. [144] proposed a cutting-in vehicle recognition functionality for adaptive cruise control utilizing a probabilistic model for situation analysis and prediction. Kasper et al. [145] enhanced this approach to 27 driving maneuvers (including merging and object following) in structured highway scenarios by using traffic scene modeling with object-oriented Bayesian networks. Lane-related coordinate systems are applied together with individual occupancy grids for all vehicles. The approach is basically designed for path prediction of surrounding vehicles and not the own one. The drawback is that lane markings are required and the network has been trained for highway speeds only. No evaluation of data from urban traffic or turn maneuvers has been presented.

Aoude et al. [146] estimate reachable sets to obtain a probabilistic description of the future paths of surrounding agents. The proposed approach combines the 'rapidly-exploring random tree reach' algorithm (RRT-Reach) and mixtures of Gaussian processes. RRT-Reach was introduced by the authors as an extension of the closed-loop RRT (CL-RRT) algorithm to compute reachable sets of moving objects in real-time. A mixture of Gaussian processes is a flexible parametric Bayesian model used to represent a distribution over trajectories. The mixture is trained using typical maneuvers learned from statistical data, and RRT-Reach utilizes samples from the Gaussian processes to grow probabilistically weighted feasible paths of the surrounding vehicles. In contrast to the work of this thesis, they incorporate data from world obstacles into trajectory predictions and show their approach only for a car-like vehicle in a simple scenario. Laugier et al. [147] predict trajectories a short period ahead using HMMs and Gaussian processes. However, their approach requires visual environment perception and telemetric data. Wiest et al. [148] infer a joint probability distribution as motion model using motion patterns that have been observed during the previous training. A standard and a variational Gaussian mixture model are compared and serve the trajectory prediction by calculation of the probability of future motion conditioned on the motion pattern of the near past. The variational Gaussian mixture model showed acceptable results; however, the approach has only been trained and tested on data from three intersections. No lane changes have been trained or tested. Thus, the robustness of the approach for a higher variety of intersections with different geometries has not been shown.

Hayashi et al. [149] address the prediction of stopping maneuvers considering the driver state. They use two different driving models dependent on the driver's state (normal or hasty based on heart rate variability). An HMM serves modeling of each driving pattern.

Huahagen et al. [150] combine a fuzzy-based rule to model basic maneuver elements and probabilistic finite-state machines to capture all possible sequences of basic elements that constitute a driving maneuver. The approach was evaluated on the recognition of turn maneuvers. Unfortunately, the proposed approach does not provide a parameterizable way for trajectory prediction and has not been evaluated for lane change maneuvers.

Tomar et al. [151] use a multi-layer perceptron (MLP) to predict the future path of a lane changing vehicle. The inputs of the network model include the relative velocity of the vehicles involved in the lane change and traveling in the passing lane, acceleration/deceleration, safety distance (time between the vehicles), and the current safety states of the vehicles. The neural network is able to predict the future states of a lane changing vehicle only in certain discrete sections of the lane change path. The approach performs well for highway speeds, but it has not been designed for urban scenarios and does not incorporate any other maneuvers.

Tsogas et al. [152] base their approach of maneuver and driver intention prediction on Dempster's and Shafer's theory of evidence. Transitions from one maneuver to another are modeled by a state diagram. However, lane marking detection is required for their solution.

Polychronopoulos [153] proposed a hierarchically structured algorithm that fuses the traffic environment data with car dynamics in order to predict the trajectory of the ego-vehicle in highway scenarios. Lane change maneuvers are detected based on the lateral offset to the road border.

Zong et al. [154] combine an ANN and an HMM in their integrated model to identify the driver's intention and to predict the maneuvering behavior of the driver. They verified their approach applying driving simulator tests on straight roads performing double-lane changes at certain vehicle speeds.

Taniguchi et al. [155] predict driving behavior using an unsupervised learning approach. They apply a semiotic, double articulation structure where they assume that unsegmented driving behavior is segmented into several short-term behaviors based on its linearity or its locality of distribution in its observed state space. They use only data that is available on CAN, but their evaluation does not include the accuracy of the resulting path prediction, especially, in case of wrong symbol prediction. They only evaluated the number of correctly predictable symbols. Furthermore, they tested their approach only on data from a few hundred meters of driving.

Liebner et al. [156] predict the driver's intent to turn or to go straight in presence of a preceding vehicle at an intersection, where they also predict the driver's intent to stop at the intersection before he completes the maneuver. They classified the driver's maneuver intention using a Bayesian classifier with uniformly distributed prior intention distribution and proposed a method to extract characteristic, desired velocity profiles that give the driver model the chance to account for turn-related deceleration. Unfortunately, their classification and prediction approach depends on the distance from the intersection center, so that it can only be used when GPS and map data are available. Furthermore, the approach has only been tested at one intersection. Therefore, its robustness is not guaranteed.

All of the described approaches show deficits and are not useful for the desired application. A novel and efficient approach is required that is robust to inaccurate sensor data, various road geometries and different drivers as well as to truck configurations with and without trailer. Furthermore, only data that is available on CAN should be required and no information from the environment perception. The new approach shall be designed for urban scenarios, where lane markings, road border information or following of the ego lane are not necessarily assumable. These features are not included in the existing approaches. It should provide the probability of the predicted paths.

One of the potential applications of the path prediction should provide information if the driver has seen an obstacle, for example a pedestrian, and is therefore able to avoid it by steering and/or braking without any system reaction. Therefore, driver interaction with possibly critical obstacles should not be considered for the prediction.

A new approach is presented using only ego-state data from CAN to predict future trajectories in a robust, computationally efficient way for the application on urban streets without lane markings. One advantage of the new approach is that it enables a pre-selection of relevant objects that should be tracked and should be provided on CAN. Another advantage is given by the fact that its results can be used for situation evaluation, e.g., in collision mitigation systems.

The proposed approach classifies maneuver types like turn, lane change and lane following based on a set of prototype trajectories using the longest-common-subsequence (LCS) method[1] in combination with a Bayesian classifier.[2] Maneuver prediction has been performed using the LCS method with a large database, sequences of higher dimension (four or six instead of two) as well as differential GPS and map data in other works, but not with single prototype trajectories and in combination with a Bayesian classifier.

5.1.2 Related Approaches for Risk Assessment

ADAS targeted at active safety have focused on preceding traffic and re-action to motorized road users in cars or trucks for a long time. These traffic participants mostly follow continuous and linear dynamic models within the relevant or considered time horizon, properties that have been exploited in [167–170]. The authors apply deterministic threat assessment and predict single trajectories for each object and use these to compute various threat measures, e.g., the metric time-to-collision (TTC) [171], pre-dicted minimum distance or predicted time to minimum distance [172]. Predicted trajectories are often based on statistical estimation methods, e.g., the Kalman filter, while in the threat assessment only a point esti-mate is used. Time metrics characterize the situation's criticality by time intervals until a predicted critical event occurs. The critical event may also be the latest moment at which an imminent collision can be avoided by a certain maneuver, i.e., braking or steering [173]. Coelingh et al. pre-sented their approach for a collision warning with full auto brake and pedestrian detection in [174]. The system is based on single trajectory predictions and the computation of the resulting TTC. It is the first sys-

---

[1]The LCS method has been applied in various disciplines. For example, it has been used for the analysis of ribonucleic acid (RNA) [157,158] or for the recognition of handwritten symbols [159]. Several algorithms have been developed to optimize the approach for the various applications [160–164].

[2]Bayesian networks have also been applied in several fields of research. A very popular application of the naive Bayesian classifier is spam filtering or text categorization. Ue-bersax [165] proposed Bayesian networks for breast cancer risk modeling in medicine, while Friedman [166] used them in biology to discover interactions between genes based on multiple expression measurements.

tem with pedestrian protection that has been on the market in passenger cars.

Polychronopoulos et al. [172] use several prediction models to represent possible future maneuvers, but the different possibilities are condensed into a single prediction without considering them as distinct possibilities. A systematic approach to choose warning thresholds in different deterministic methods is given in [175], while Jansson [176] presented an overview of different deterministic threat assessment methods.

Two types of uncertainties are relevant for criticality assessment: First, the prediction uncertainty that is present, since the system can only make assumptions about the future actions and plausible behavior of the road users. Second, the inaccuracies of the sensors perceiving the current situation and providing the attributes of the detected objects.

Berthelot et al. [177] determine a collision probability ($P(\text{TTC} \in \mathbb{R})$) that relies on a non-Euclidean distance function for extended objects and its statistical linearization via the unscented transformation. In [178], they extend their work and compute the probability distribution of the TTC induced by an uncertain system input.

Rodemerk et al. [179] develop a general criticality criterion for driving situations where the areas that objects could cover and the amount of overlap determine the collision probability depending on the calculation time. The model for trajectory prediction assumes constant lateral and longitudinal accelerations. Lefèvre et al. [180] estimate the collision risk at road intersections with a dynamic Bayesian network, where they compare driver intention and expectation.

Ferrara et al. [181] investigate the possibility of reducing the number of accidents involving pedestrians or other vulnerable road users, like cyclists and motorcyclists. They use a supervisor scheme to choose the appropriate current control mode for each controlled vehicle and manage the switches among low-level controllers. They consider the sliding mode control methodology suitable to deal with uncertainties and disturbances.

Wakim et al. [182] applied Monte-Carlo simulations to predict collision probabilities with pedestrians, where they model pedestrian dynamics with four states (static, walk, jog and run) as truncated Gaussians using an HMM. Large [183] applies a cluster-based technique to learn motion patterns using pairwise clustering. Pedestrian motion is predicted using the cluster mean value. The resulting trajectory is used to estimate the collision probability of the pedestrian with a vehicle. Gandhi summarized additional approaches for pedestrian collision avoidance in [112].

Westenhofen et al. [83] calculate the collision risk between vehicles and pedestrians based on the vehicles' driving tubes that are determined by means of the linear bicycle model taking into account physical constraints. A pedestrian movement area is determined using a physiological model that returns the acceleration and turning capability of a human for a specific walking speed. They compare the intersection between the movement area of the vehicle and the movement area of the pedestrian to the total movement area of the pedestrian to determine the collision risk. The transponder-based system is advantageous when object occlusion has to be handled; however, system tests with artificially generated dummy scenarios showed a significant number of false system reactions.

Smith and Källhammer [184] used location and motion information of pedestrians to identify when drivers would accept system alerts. The results are based on environment information that is not provided by the sensors, such as exact and robust lane information, road border elements between the road and the sidewalks. Thus, the approach cannot be applied here.

Monte-Carlo techniques [185–188] incorporate the option to consider complex, highly non-linear vehicle dynamics, but the techniques usually require many simulation runs for suitably capturing the possible evolutions. Nevertheless, these methods cannot cover all possible behaviors in principle.

Additional approaches for stochastic threat assessment including other disciplines are given in [71, 176, 185–192]. Broadhurst el al. [191] aim to produce collision-free trajectories and proposed a method for a prediction and path planning framework for road safety analysis. They consider combined actions and interactions of all detected objects and objects that may enter or leave the scene. Their map-based approach models object interaction using game theory, where all object decisions are enumerated in a decision tree. The computation of the car's trajectory uses the given initial state of the car and a series of control inputs (acceleration and steering). They already incorporate knowledge about sensor uncertainty due to incorrectly classified objects, imprecise measurements, missing detections or grouping multiple objects into a single detection result. In later works, Broadhurst el al. [185] and Eidehall et al. [186] consider many (thousand) different possibilities for all objects by trying to approximate the true probability distribution of the predictions using Monte-Carlo sampling. The current states of all objects are assumed as given stochastic variables. They suggest a driver preference distribution to model the aspects distance to intended path, deviation from desired velocity, longitudinal acceleration and steering angle or lateral acceleration.

The cost of a maneuver of a single object for the entire time interval is computed, where weights can balance the costs between velocity or path deviation. In that work, they did not consider sensor uncertainty and used only synthetic data. They were interested in determining which future control inputs are safe and which are dangerous. Eidehall et al. [186] consider road friction as the limiting factor for longitudinal input at low speeds and engine power at high speeds. The dynamics of pedestrians and bicycles are modeled assuming constant accelerations.

Danielsson and Eidehall [187, 192] use Monte-Carlo simulation to find threats in a road scene. They adapt their model for calculations in coordinate system that is aligned to a curved road. They use time intervals for their computations but they do not consider position intervals. Their algorithm was only tested with recorded data and they did not show that the algorithm was running in real-time with the required number of samples. Even though the number of samples in the final distribution is high, many of the samples are based on the same parents, limiting the statistical variation. The results are not repeatable as the sample generation is random, which is different in this work due to a pre-computed transition matrix.

Stochastic verification techniques have been investigated to increase air traffic safety [190]. Althoff [71] presented a scalable approach for real-time prediction of potentially hazardous situations based on stochastic reachable sets of traffic participants by defining discrete actions, e.g., acceleration or braking. He took into account multiple road users. However, the work concentrates on road users like cars and trucks that follow the road according to defined dynamic models. The algorithm is designed for safety verification of planned paths in an autonomous car.

For the development of ADAS in commercial vehicles, the human driver has to be taken into account as the controlling instance. Therefore, the planned trajectory of the ego vehicle has to be estimated from signals (e.g., acceleration and steering angle). Thus, the focus of this work is not on path planning but on checking whether or not a collision with another road user is probable to occur and on deciding if an active system reaction should be triggered. It is assumed that drivers do not recognize the vulnerable road users, so that no interaction between the driver and the other road users shall be taken into account. The consideration of possible false detections from the environment perception is very important, since these might lead to a more critical situation evaluation and therefore trigger a false system reaction. This would make the application more dangerous from the perspective of functional safety, since the application is compared to a system not equipped with this functionality.

The newly-developed approach considers different prototype paths that are followed with a certain probability. In contrast to [71], pedestrians are explicitly modeled in this work. As pedestrians do not necessarily follow the lane, possible prototype paths on which they will walk are defined, and path probabilities are assigned according to the positions and velocities of previous measurement cycles.

5.2 Maneuver Classification and Trajectory Prediction

The principle of the novel approach for maneuver classification and trajectory prediction is visualized in Figure 5.1 and can be summarized as follows. Data from CAN is recorded, filtered and buffered to obtain in-

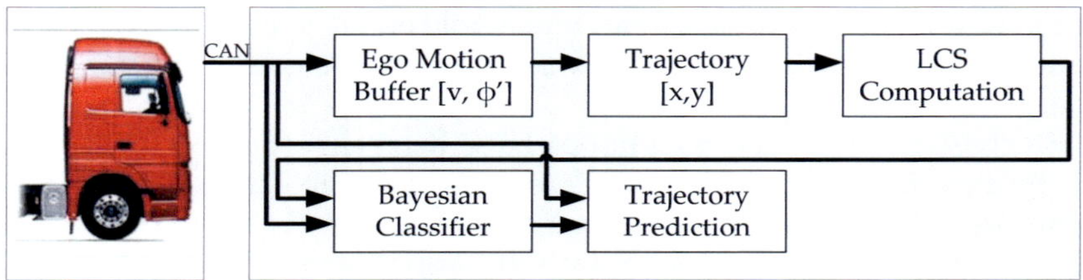

Figure 5.1 Principle of the approach for maneuver classification and trajectory prediction.

formation about the history of the trajectory (ego motion buffer). The past trajectory sequence is compared to different so-called prototype trajectories using the LCS method providing a first classification indicator. The meaning and the determination of the prototype trajectory is explained below. The result of this primary classification step is fed to a Bayesian classifier together with other signals to estimate the maneuver class that is intended by the driver, where the maneuvers turn, lane change and lane following are possible outcomes. Next, the prediction is performed using information from the prototype trajectory of the estimated maneuver class. The development of the approach for maneuver classification and trajectory prediction has been supported by the supervised student research project of Matheus Dambros [223]. First results have been presented in [208].

A trajectory $\mathbf{T}_{(\cdot,\cdot)}$ is defined as a sequence of N_{tr} pairs containing odometric data like velocities v and yaw rates $\dot{\phi}$ or xy-positions in Cartesian space with associated timestamps t_k:

$$\mathbf{T}_{v,\dot{\phi}} = ((v_0, \dot{\phi}_0), t_0), \ldots, ((v_k, \dot{\phi}_k), t_k), ((v_{N_{tr}-1}, \dot{\phi}_{N_{tr}-1}), t_{N_{tr}-1}) \tag{5.1}$$

$$\mathbf{T}_{x,y} = ((x_0, y_0), t_0), \ldots, ((x_k, y_k), t_k), ((x_{N_{tr}-1}, y_{N_{tr}-1}), t_{N_{tr}-1}) \tag{5.2}$$

where $t_k < t_{k+1}$ for $k = 0, \ldots, N_{tr} - 1$. The first form of trajectory representation implies the advantage that a rotation of the trajectory is represented by a subtraction of the rotation angle from the yaw angle.

The number of trajectory sequence elements can either be limited by a fixed number of samples N_{tr}, a maximum length l_{tr} in meters or by its duration $\Delta t_d = t_{N_{tr}-1} - t_0$. The number of elements N_{tr} can vary in the two latter cases, since the measurements are not always taken at equidistant points in time. Wiest et al. [148] applied a Chebyshev decomposition on the components of the trajectory to obtain a uniform representation of the trajectory and used the resulting Chebyshev coefficients as input features for their model. However, evaluations showed that simple linear signal interpolation suffices here. Moreover, the interpolation of the signal values makes the strongly quantized yaw rate signal smoother.

The trajectory representation does not depend on a specific sensor type and may result from several different sensor types. Other authors [60,65] determined the motion of the ego vehicle using axle speeds and single wheel speeds as well as data from yaw rate sensors. Unfortunately, the available test truck does not provide single wheel speeds nor GPS information on CAN so that only the speed information from the axles and the data from the yaw rate sensor can be utilized. The quantized values of the yaw rate sensor have to be corrected using a temperature depending factor and offset. This circumstance makes it hard to always correctly record and predict the ego vehicle's path but requires a robust approach for trajectory prediction that does not depend on the sensor's performance and accuracy.

The filtered yaw rate data and the rear-axle speed information of the ego vehicle are written to a buffer that contains the data from a duration of three seconds. Samples with zero speed, e.g., due to waiting times at red lights, are cut and excluded from the data since they do not provide any informational benefit. Next, the data is converted to path positions $[x_e, y_e]^T$ in Cartesian space by integration according to the following equations

$$x_{e,k} = x_{e,k-1} + v_{e,k} \cdot \sin(\phi_{e,k}) \cdot \Delta t, \tag{5.3}$$

$$y_{e,k} = y_{e,k-1} + v_{e,k} \cdot \cos(\phi_{e,k}) \cdot \Delta t, \tag{5.4}$$

$$\phi_{e,k} = \phi_{e,k-1} + \dot{\phi}_{e,k} \cdot \Delta t, \tag{5.5}$$

where $\dot{\phi}_{e,k}$ is the yaw rate at time point k and $v_{e,k}$ the corresponding tangential ego speed. The oldest sample contained in the trajectory buffer defines the origin of the Cartesian coordinate system. This type of trajectory representation depends on the orientation, the scale and the start position.

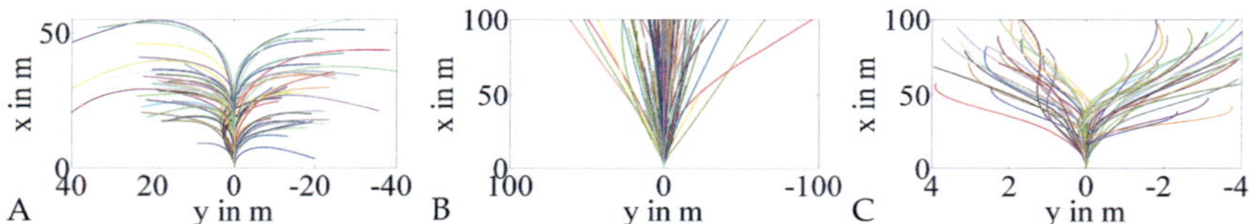

Figure 5.2   Various trajectory representations recorded on various streets and intersections: **A)** turn maneuvers; **B)** lane following maneuvers; **C)** lane change maneuvers.

Figure 5.2 shows different representations of the maneuver classes turn, lane change and lane following in Cartesian space, where no distinction is made between maneuvers to the left and to the right. The labeled maneuvers have been recorded on various streets and intersections. The figure illustrates that a maneuver classification approach using only data from single intersections and streets is not expected to perform very robustly.

However, one contribution of this work is the classification of maneuvers, although their representations might be manifold. There are many intersections where the streets do not cross perpendicularly, so that the inner angle of a turn maneuver might be very large and the corresponding trajectory looks similar to lane following on a curved street. Therefore, a distinction between turn and lane following for the manual labeling is made as follows: Maneuvers including inner angles with less than $135°$ are considered as turn maneuvers and all the ones with larger inner angles as lane following. The special challenge implied by lane change maneuvers on curved streets will be discussed later.

Since the trajectory should be classified, it has to be comparable to other trajectories of the same class requiring an efficient representation that is independent from translation and rotation. This form can either be obtained directly from the trajectory $\mathbf{T}_{v,\dot{\phi}}$ containing speed and yaw rate or from the Cartesian representation $\mathbf{T}_{x,y}$. The explanation is based on the Cartesian representation here for an easier illustration, see Figure 5.3. The

Figure 5.3 Trajectory pieces in Cartesian space and computation of the representation in arc lengths and angles. **A)** current position on trajectory with used history data and predicted path; **B)** arc lengths and angles.

trajectory is transformed to a representation based on arc lengths and angles. The length of a line segment between two successive measurement points k and $k + 1$ represents an arc element l_k:

$$\mathbf{l}_k = \begin{bmatrix} x_{k+1} - x_k \\ y_{k+1} - y_k \end{bmatrix}, \tag{5.6}$$

$$l_k = \|\mathbf{l}_k\| = \sqrt{(x_{k+1} - x_k)^2 + (y_{k+1} - y_k)^2}. \tag{5.7}$$

The computation of the counter-clockwise angles between these line segments leads to a sequence of angles. The calculation of the cross product and the scalar product between two successive line segments $\mathbf{l}_k$ and $\mathbf{l}_{k+1}$ supports the computation of angle θ_k between these two line segments:

$$\theta_k = \mathrm{sign}(\mathbf{l}_k \times \mathbf{l}_{k+1}) \cdot \left(\pi + \arccos\left(\frac{\mathbf{l}_k \cdot \mathbf{l}_{k+1}^{\mathrm{T}}}{l_k \cdot l_{k+1}} \right) \right). \tag{5.8}$$

The value π may be subtracted from the angle elements, since it is added to all angle elements and does not provide any benefit. Furthermore, the mean of the total angle sequence is subtracted from each element to make the sequence independent from driving on a straight or a curved street. The absolute values of the resulting angle sequence can be used to enable a maneuver type classification that is independent of the direction.

The obtained trajectory representation $\mathbf{T}_{l,\theta}$ contains concatenated pairs of arc lengths and angles:

$$\mathbf{T}_{l,\theta} = (l_0, \theta_0), \ldots, (l_k, \theta_k), \ldots, (l_{N_{tr}-2}, \theta_{N_{tr}-2}). \tag{5.9}$$

An example of a recorded trajectory in Cartesian space and the computed angle and arc length sequences over time can be found in Figure 5.4. According to own evaluations, the proposed method showed the

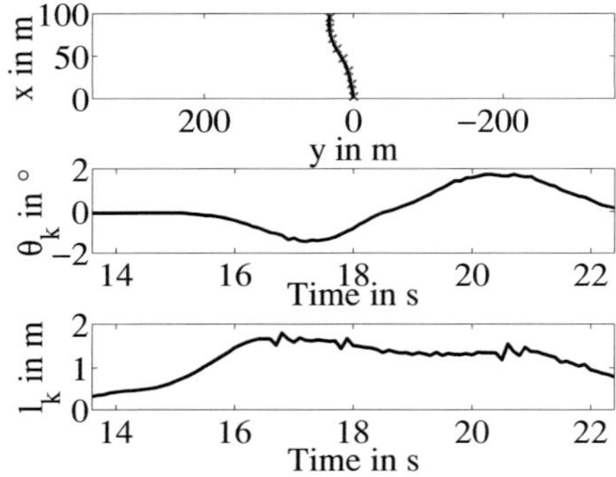

Figure 5.4   Example of a trajectory in Cartesian space based on a sample every 100 ms and the corresponding angle or arc sequences over time: The sequences between red crosses correspond to one second.

best properties considering stability, robustness and computational effort compared to another method where an angle sequence was obtained from angles between lines that connected the center of mass of the considered trajectory sequence to the trajectory sample points. Moreover, a sampling time of 0.1 seconds showed better LCS-based classification results than using sampling times of 0.05 seconds.

5.2.1 Longest Common Subsequence

The obtained angle sequence of the trajectory is compared to prototype trajectories using the LCS method in the next step. Each prototype trajectory represents one maneuver class — a turn, lane change or lane following maneuver. Data observation demonstrates that specific, characteristic functions describe the yaw rate signals of these different maneuver classes. While the yaw rate signal of the maneuver class lane following $\dot{\phi}_{lf} = \dot{\phi}_e$ shows a constant behavior, the yaw rate signal of turn maneuvers can be represented by a Gaussian curve $\dot{\phi}_{turn}$:

$$\dot{\phi}_{turn,k} = a_{turn} \cdot \exp\left(-\left(\frac{t_{p,k} - b_{turn}}{c_{turn}}\right)^2\right) + d_{turn} , \qquad (5.10)$$

where $a_{turn} = 0.01384$, $b_{turn} = 0.00549$, $c_{turn} = 2.479$ and $d_{turn} = -0.002$ are parameters that have been estimated before from clusters of recorded and labeled yaw rate signals. The time $t_{p,k}$ with $-\tau_{p,turn}/2 \leq t_{p,k} \leq \tau_{p,turn}/2$ represents the relative timestamps of the prototype trajectory in seconds, where $\tau_{p,turn} = 9$ s for turn maneuvers. The duration of a lane following maneuver is set to 9 seconds as well.

Lane change trajectories imply a yaw rate signal that looks similar to a sine or cosine curve. Nevertheless, an approximation by a superposition of three exponential functions and a constant showed better results for the approximation of the prototype yaw rate signal $\dot{\phi}_{lc}$:

$$\dot{\phi}_{lc,k} = a_{lc,1} \cdot \exp\left(-\left(\frac{t_{p,k} - b_{lc,1}}{c_{lc,1}}\right)^2\right) +$$

$$a_{lc,2} \cdot \exp\left(-\left(\frac{t_{p,k} - b_{lc,2}}{c_{lc,2}}\right)^2\right) +$$

$$a_{lc,3} \cdot \exp\left(-\left(\frac{t_{p,k} - b_{lc,3}}{c_{lc,3}}\right)^2\right) + d_{lc} , \qquad (5.11)$$

where $-\tau_{p,lc}/2 \leq t_{p,k} \leq \tau_{p,lc}/2$ and $\tau_{p,lc} = 7.4$ s. Appropriate values for the parameters have been found for $a_{lc,1} = 0.001836$, $b_{lc,1} = -1.118$, $c_{lc,1} = 1.424$, $a_{lc,2} = 0.001391$, $b_{lc,2} = 1.314$, $c_{lc,2} = 1.186$, $a_{lc,3} = -0.00133$, $b_{lc,3} = -0.07651$, $c_{lc,3} = 0.714$ and $d_{lc} = -0.0005$. Figure 5.5 illustrates the described characteristic and shows the unfiltered yaw rate signals together with its function-based approximations.

A combination of the characteristic yaw rate signals with the current speed of the vehicles leads to a prototype trajectory $\mathbf{T}_{v,\dot{\phi},p}$ for each maneuver class. Finally, trajectory representations in Cartesian space $\mathbf{T}_{x,y,p}$ and in form of arc and angle $\mathbf{T}_{l,\theta,p}$ can be obtained as described above and can be used for comparison with the history of the previous trajectory piece $\mathbf{T}_{l,\theta,e-}$ to obtain a similarity measure. The angle sequences of both trajectories are interpreted as symbol sequences $\mathbf{T}_{\theta,p}$ and $\mathbf{T}_{\theta,e-}$.

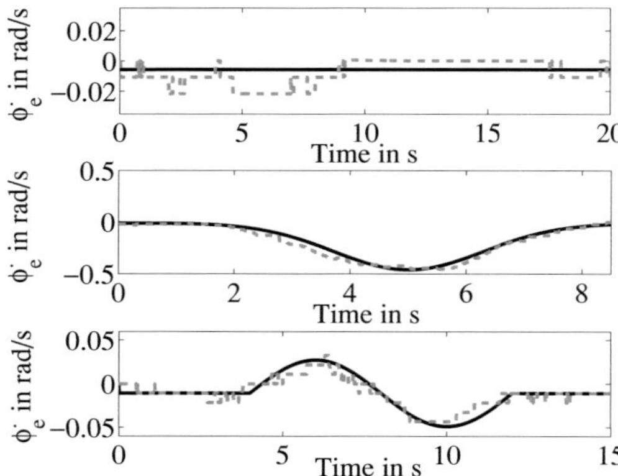

Figure 5.5 Yaw rate $\dot{\phi}_e$ of recorded maneuvers (dashed curves) lane following, turn and lane change (from top to bottom) and corresponding prototype yaw rates (solid curves).

The LCS method compares both sequences by looking for same symbol orders in both sequences. The output is the length of the LCS. The common symbol sequence itself can be computed as well, but at much higher computational effort. The LCS method has the advantage that symbols do not have to follow directly after each other making more robust, e.g., to outliers, than a similar method called the longest-common-substring method.

The sampling time has been set to 100 ms and the used driven trajectory piece $\mathbf{T}_{l,\theta,e-}$ has a temporal length of 3 s, which leads to a symbol number of $N_{e-} = 30$. Based on the maneuver durations, the symbol sequence of the lane change prototype $\mathbf{T}_{\theta,\mathrm{lc}}$ contains $M_{\mathrm{lc}} = 74$ symbols, while the symbol numbers of the other maneuver classes are $M_{\mathrm{turn}} = 90$ and $M_{\mathrm{lf}} = 90$.

A dynamic programming approach has been chosen for the implementation of the LCS value computation, since it avoids double computations. The procedure may be divided into the computation of the length of the LCS and the determination of the sequence itself. Only the length of the LCS is relevant here, so that the computationally demanding part may be neglected. The computation time and memory requirements can be reduced by an optimization step. The dynamic programming algorithm builds a matrix $\mathbf{C}$ of dimension $N_{e-} \times M_{\mathrm{p}}$, where M_{p} represents the number of symbols of the corresponding prototype trajectory. The matrix is initialized with zeros. The algorithm runs through lines and columns checking if the absolute difference of the elements i of the first sequence $\mathbf{T}_{\theta,p}$ and the elements j of the second sequence $\mathbf{T}_{\theta,e-}$ are below a given threshold $\epsilon_T(i)$. In this case, the value of $\mathbf{C}_{i,j} + 1$ is assigned to $\mathbf{C}_{i+1,j+1}$. Otherwise, the maximum of the adjacent values $\mathbf{C}_{i,j+1}$ and $\mathbf{C}_{i+1,j}$ are set for the value of $\mathbf{C}_{i+1,j+1}$:

$$\mathbf{C}_{i+1,j+1} = \begin{cases} \mathbf{C}_{i,j} + 1, & \text{if } \left|\mathbf{T}_{\theta,p}(i) - \mathbf{T}_{\theta,e-}(j)\right| \leq \epsilon_T(i) \\ & \wedge \; |i - j| < N_{\text{tol}} \\ \max(\mathbf{C}_{i,j+1}, \mathbf{C}_{i+1,j}), & \text{otherwise,} \end{cases} \tag{5.12}$$

where N_{tol} limits the number of symbols that may lay between similar symbols. The element $\mathbf{C}_{M_p,N_{e-}} = \mathbf{C}_{i,j}$ with $i = M_p$ and $j = N_{e-}$ corresponds to the absolute length of the LCS of the two symbol sequences. The values of $\epsilon_T(i)$ depend on the maneuver class, since the symbols in the lane change sequences show a smaller range than the symbols in the turn sequences. Signal clusters of the corresponding maneuvers serve the computation of the threshold value, where the standard deviations from the mean values appeared as proper measure. A normalization step is performed next. It depends on the number of used symbols, so that all LCS values are within a range of 0 to 100 and can be used for further computations:

$$n_{\text{lcs}} = \frac{\mathbf{C}_{M_p,N_{e-}}}{\min(M_p, N_{e-})} \cdot 100. \tag{5.13}$$

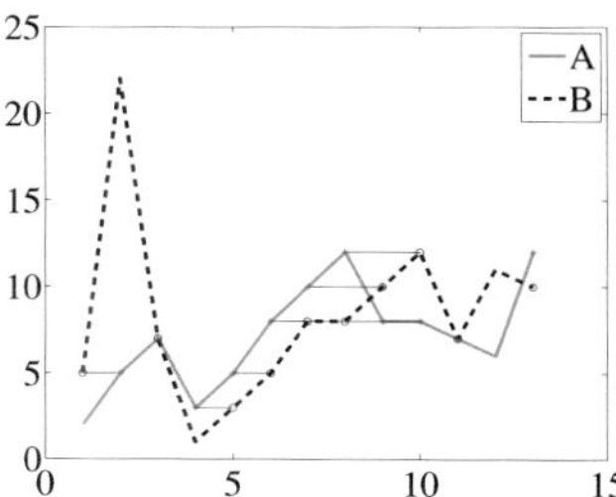

Figure 5.6 Visualization of the LCS principle: The length of longest common subsequence of the sequences A and B is $n_{\text{lcs}} = 8$ when $\epsilon_T = 0$.

Figure 5.6 shows a short example for two sequences and their common symbols. One obtains one LCS value n_{lcs} for each comparison with a prototype trajectory ($n_{\text{lcs,lf}}$, $n_{\text{lcs,lc}}$ and $n_{\text{lcs,turn}}$).

5.2.2 Naive Bayesian Classifier

The Bayesian classifier is a statistic classifier that aims to maximize the posterior probability of a class given its features. It has a decision function that is maximum square.

Here, the features are based on the ego vehicle's speed v_e, the acceleration a_e, the steering wheel angle $\phi_{steer,e}$, the yaw rate and the LCS results n_{lcs}. Additional features that will be explained later in this section are computed based on these signals. Signals like kickdown and signals from turning lights, the brake pedal or the throttle have been investigated but provide only little benefit. Possible classes are the maneuvers turn, lane change and lane following, so that the conditional probability $P(\text{mv}|n_{lcs,lf}, \ldots, v_e)$ should be maximized. The Bayesian formula serves this task:

$$P(\text{mv}|n_{lcs,lf}, \ldots, v_e) = \frac{P(\text{mv})P(n_{lcs,lf}, \ldots, v_e|\text{mv})}{P(n_{lcs,lf}, \ldots, v_e)} \tag{5.14}$$

$$= \frac{P(\text{mv}, n_{lcs,lf}, \ldots, v_e)}{\text{const}}, \tag{5.15}$$

where $P(\text{mv})$ is the prior probability of the maneuver class. The numerator of Equation 5.15 is the joint probability model $P(\text{mv}, n_{lcs,lf}, \ldots, v_e)$ and can be rewritten using the chain rule. This allows that any member of the joint distribution can be calculated using only conditional probabilities. The Bayesian classifier favors classes that occur more often, which is the optimal strategy if the risk of a false classification is to be minimized. If the prior probability is unknown, it could be assumed uniformly distributed, resulting in a maximum-likelihood classifier.

In this work, two approaches for the determination of the prior probability have been chosen. For the first approach, information about representative routes for commercial vehicles in the urban and sub-urban environment was obtained from a logistics company. These routes have been driven by different drivers with and without trailer and data has been recorded for about four hours. Then this data has been labeled by manually assigning a maneuver class to each sample. The fraction of a certain maneuver class in the data has been set as corresponding prior probability. The prior probabilities have been computed to $P(\text{lf}) = 0.87$, $P(\text{turn}) = 0.09$ and $P(\text{lc}) = 0.04$. The second approach trained the prior probability of the maneuver classes based on the classification result. The path prediction of a lane following maneuver is assumed to show the lowest risk under presence of uncertainty compared to the maneuvers

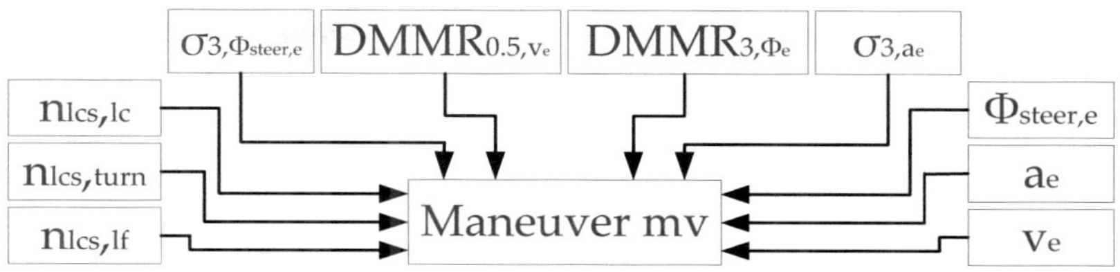

Figure 5.7 Bayesian classifier.

turn and lane change. Therefore, the prior probability $P(\text{lf})$ is set so high that quasi no maneuver is falsely classified as lane change or turn maneuver. The obtained prior probabilities are similar to the first approach ($P(\text{lf}) = 0.865$, $P(\text{turn}) = 0.095$ and $P(\text{lc}) = 0.04$).

Determination of all conditional probabilities makes solving the problem very complex and requires many parameters. Therefore, the *naive conditional independence assumption* is applied stating that each feature is independent from any other feature, e.g., $P(v_e|\text{mv}, \phi_{\text{steer,e}}) = P(v_e|\text{mv})$. The resulting classifier is called *naive Bayesian classifier* due to the conditional independence assumption and can be formulated by:

$$P(\text{mv}|n_{\text{lcs,lf}}, \dots, v_e) = P(\text{mv}) \cdot \frac{P(n_{\text{lcs,lf}}|\text{mv}) \cdot \dots \cdot P(v_e|\text{mv})}{\text{const}}. \quad (5.16)$$

Conditional independence is not fulfilled for some of the features. For example, n_{lcs} depends on the speed v_e and the yaw rate that is related to the steering wheel angle $\phi_{\text{steer,e}}$ under normal driving conditions. However, naive Bayesian classifiers have worked quite well in practice, although they work with over-simplified assumptions.[3] The naive Bayesian classifier requires a smaller amount of training data than many other classifiers to estimate the parameters required for an accurate classification, which makes it especially attractive when data sets contain many classes and many features. The number of parameters that is required for the classifier depends on the number of classes (3 here), the number of features (10 here) and the number of parameters it takes to describe the conditional probability of a feature given its class (e.g., 1 in case of Bernoulli variables, 2 in case of Gaussian distributions).

---

[3]Zhang [193] showed some theoretical reasons for the apparently high effectiveness of naive Bayesian classifiers.

A distribution for each feature and its parameters is estimated using a training data set. Gaussian kernels[4] model the feature distribution of the speed v_e. The LCS features n_{lcs} are represented by bounded Gaussian kernel distributions, where the distribution is bounded between 0 and 100. The steering wheel angle $\phi_{steer,e}$ and the acceleration a_e follow simple Gaussian distributions. Furthermore, some special features have been created from the signals. The standard deviation of the acceleration σ_{3,a_e} and the standard deviation of the steering wheel angle $\sigma_{3,\phi_{steer,e}}$ from the measurements of the last three seconds have been computed as additional features to invoke the variable dynamic behavior of a maneuver. The maximum speed range of the last 500 ms $DMMR_{0.5,v_e}$ and the maximum range of the angle ϕ_e within the last three seconds $DMMR_{3,\phi_e}$, both related to the corresponding time interval, serve as additional input of the classifier, where DMMR is the abbreviation for differential maximum minimum range. Gaussian kernel distributions model the last four features σ_{3,a_e}, $\sigma3$, $\phi_{steer,e}$, $DMMR_{0.5,v_e}$ and $DMMR_{3,\phi_e}$.
The kernel density estimator function is given by:

$$f_{kernel}(N_{sample}, b, x) = \frac{1}{N_{sample}b} \sum_{i=1}^{N_{sample}} K\left(\frac{x - x_i}{b}\right), \qquad (5.17)$$

where N_{sample} is the number of available samples, b the kernel bandwidth, x_i are the values of the single samples and $K(\cdot)$ is the kernel function, which is a normal distribution $\mathcal{N}(\mu, \sigma) = \mathcal{N}(\mu, b)$ here. The bandwidth of the kernel is a free parameter that has to be optimized to obtain the best density approximation. The kernel bandwidth has been estimated using kernel density estimation based on [195]. Plots of the resulting kernel densities can be found together with the classification results in Section 6.2. Figure 5.7 illustrates the resulting Bayesian classifier.

When the Bayesian classifier is applied to classify a maneuver based on new samples, the posterior probability of that sample belonging to each class is computed by inserting the feature into the corresponding distribution function with class-dependent parameters. The largest posterior probability determines the class that the new sample is expected to belong to according to the classifier.

---

[4]In statistics: Kernel density estimation (KDE) is a non-parametric way to estimate the PDF of a random variable. KDE is a data smoothing problem where inferences about the population are made based on a finite data sample [194]. If a Gaussian function is chosen as kernel function, one refers to it as Gaussian kernel.

5.2.3 Path Prediction

Two approaches have been developed for the ego vehicle's trajectory prediction. The first approach uses the computed prototype trajectories and the obtained maneuver class as basis for the prediction. The estimated maneuver class determines which prototype trajectory shall be chosen. Then, the yaw rate of this prototype trajectory is combined with the current ego vehicle speed and acceleration, so that a new trajectory $\mathbf{T}_{v,\dot{\phi}+}$ results. A part of this trajectory can be taken as the estimated future trajectory. In case of estimated lane following, the yaw rate is adopted to the current situation by setting it constant to the current yaw rate. Thereby, a prediction on streets with different curvatures is possible.

In case of lane change and turn maneuvers, it is important to know at which point of the trajectory the ego vehicle is currently located to choose the proper part of the newly generated trajectory for prediction. This is possible using the LCS computation. The previously driven trajectory piece $\mathbf{T}_{l,\theta,e-}$ is compared to subsequences of the sequence $\mathbf{T}_{v,\dot{\phi}+}$ that have the same duration as $\mathbf{T}_{l,\theta,e-}$ and lead to the same number of symbols N_{e-}. The subsequences are obtained by shifting a window of length N_{e-} over the symbol sequence of $\mathbf{T}_{v,\dot{\phi}+}$. The subsequence that shows the highest congruence with the previously driven trajectory piece $\mathbf{T}_{l,\theta,e-}$ determines where the predicted trajectory starts. Let i_T be the start index of the corresponding subsequence related to the total sequence. Then, the predicted trajectory is the subsequence of $\mathbf{T}_{v,\dot{\phi}+}$ that starts at index $i_\mathrm{T} + N_{e-} + 1$.

The second approach is based on the trajectory symmetry around the reference points of lane change maneuvers and turn maneuvers. Apex (turn) or inflection point (lane change) of the yaw rate signal from CAN are determinable using a few heuristics and the evolution of the yaw rate signal is axially symmetric around the apex or roughly symmetric with respect to the inflection point, respectively. This knowledge is exploited for the yaw rate prediction and thus, the path prediction. A constant speed is used for prediction of lane change maneuvers, whereas the assumption of constant acceleration is made for the prediction of turn maneuvers. The trajectory $\mathbf{T}_{v,\dot{\phi}+}$ is computed based on a constant yaw rate until the reference point is passed. Thus, the prediction is the same as for lane following in this time interval. When the algorithm detects that the reference point has been passed, the point reflected (lane change) or line reflected (turn) yaw rate is used for prediction.

The second approach showed more accurate results for the prediction due to a more flexible adaptation to differing road geometries. Furthermore, the second approach is more efficient, since the current maneuver section (before or after reference point) can be computed using a few heuristics and no computationally expensive LCS computation has to be applied. Therefore, only the second approach is considered for path prediction in the following. The predicted path is a prerequisite for the next step, where a new approach is presented that aims to assess the criticality of a situation.

5.3 Risk Assessment

It is important to have some measure of criticality for evaluating whether or not the selected control input of the driver is appropriate and if a corrective action should be taken before a safety-critical application intervenes actively into the system. The additional warning phase makes the situation evaluation more challenging, since the prediction needs to look about three seconds into the future. A pedestrian can cross the driving corridor (width of 3.5 m) with a constant speed of 1.5 m/s in only 2.3 s.

The dynamics of pedestrians are highly volatile and do often not follow any known continuous or linear dynamic model. Due to the high level of inaccuracy and uncertainty caused by measured data and barely known future behavior of the road users, it is insufficient to consider only single maneuvers in order to decide if a detected pedestrian should trigger a system reaction. Multiple possibilities have to be evaluated.

Figure 5.8 summarizes the principle of the proposed approach. In an offline step, the state space (position and speed) and the input space (acceleration) are discretized to cell arrays representing uncertainty in the position, the speed, and the acceleration of the road users. Transition matrices for time points and time intervals are computed for each road user class based on class-dependent parameters by using Monte-Carlo simulation and Markov chain abstraction. In the online application, the class-dependent transition matrices are loaded for the ego vehicle and each detected road user. A stochastic reachable state set is computed for each road user and a certain prediction interval using these transition matrices as well as previously predicted paths. Intersection of the reachable positions of the ego vehicle and a detected road user returns the conditioned crash probability. In the crash probability, the approach takes into account uncertainty in the future behavior of the road users and in the real existence of detected road users. Prototype paths are assigned to the ego vehicle and the pedestrians based on the observations from previous

<table>
<tr>
<td valign="top">

Offline approach for different classes of road users:

- Definition and discretization of state space and input space
- Definition of time increment for time steps and time intervals
- Monte-Carlo simulation for all initial state sets, input sets and parameters sets with state-dependent dynamic equations using a hybrid automaton
- Transition matrices for time points and time intervals using Markov chain abstraction

</td>
<td valign="top">

Online approach for all present road users:

- Interval sets of initial states from environment perception
- Estimation of paths and their probabilities
- Prediction of input probabilities
- Reachable sets of speed and position along the paths for time points and time intervals applying transition matrices dependent on road user class
- Reachable positions from reachable sets and estimated path deviations
- Intersection of reachable positions of ego vehicle and other road users

</td>
</tr>
</table>

Figure 5.8 Overview of the risk assessment approach: The offline computation of transition matrices for the different road user classes contributes to an efficient online computation of the reachable sets of the road users.

measurement cycles. A probability is assigned to each path. An efficient online algorithm computes the partial, conditioned collision probability of the ego vehicle with a pedestrian under consideration of the path probabilities by intersecting their stochastic reachable sets.

5.3.1 Study on Pedestrian Behavior

Typical pedestrian paths, the pedestrians' dynamics along these paths, as well as the maximum lateral deviation from the initial path have to be determined. This requires the analysis of pedestrian motion behavior.

Parameters for pedestrian dynamics from the literature, such as acceleration and maximum speed, have been enumerated in Section 2.3.2. The literature rather provides general information about pedestrian dynamics than pedestrian behavior in traffic in specific situations. It does not describe the dependency of the pedestrian behavior on the location and the time of the day, both of which are important dependencies.

Therefore, an exploratory study including these features was carried out in course of this work. Its goal was to gain additional information and to validate the results from the literature. The study results have been obtained during the supervised Bachelor thesis of Isabel Thomas [224].

The central questions to be answered were: How do pedestrians choose their paths and speeds across the street at different types of pedestrian crossings? Which factors influence their behavior, such as interaction with other pedestrians or attached accessories?

Hypotheses have been developed based on pedestrian observation in the daily life before measurements have been recorded in traffic using the available sensors. Different times of the day have been chosen for the recordings. The truck was stopped or parked on different urban streets with pedestrians and sidewalks — close to zebra crosswalks, crosswalks at traffic lights, crosswalks without specific infrastructure, crossings, or roundabouts. Pedestrian tracks were recorded and video data from several hours served the comparison of similar situations, such as crossing behavior at zebras with or without other pedestrians on the street. As children younger than about 7 years appeared only rarely in the recorded scenarios, the analysis does not include their behavior.

The major findings of the study are presented in the following: Usually, pedestrians try to find the shortest path across the street. However, pedestrians will sometimes cross the street in a diagonal manner if the destination is located laterally from the shortest path and there are no or only few other pedestrians on the street. The maximum observed relative deviation from the shortest path was 0.25 m/m, where zebra widths and orientations as well as street widths served as reference. Furthermore, deviations from the shortest path may be induced by slower or oncoming pedestrians on the street. Pedestrians will rather take a detour to avoid collisions with other pedestrians than adjust their speed. Pedestrian speeds on the street stayed about constant, except when the pedestrians were waiting for other slower pedestrians they seemed to know (induced deceleration) or the traffic light switched from green to red (induced acceleration). If there are still several pedestrians on the street, a change in the traffic light will not lead to an acceleration. A related behavior can be observed at zebra crosswalks. If a group crosses the street and a car approaches the zebra crosswalk, the last individuals of the group will tend to accelerate, while the others will keep their speed constant. Single individuals with the aim to cross the street will usually decelerate at the edge of the street and look around before they cross if there are not yet any other pedestrians on the street. This is different to the case with other, closely preceding pedestrians on the street. The individuals then keep on walking without observable deceleration. Pedestrians will rather follow piecewise straight paths (polygons) than walk on curved paths if their motion is not limited by other road users and they walk parallel to the street on sidewalks. However, the parallel orientation to the street is not true if the street bypasses a square.

The results from the literature regarding pedestrian speed and acceleration could be confirmed. Additional information was gathered. Groups with about 1 m/s and older individuals with about 0.9 m/s on average move more slowly than single and younger individuals (1.5 m/s on average) and their accelerations show also lower values. People with an estimated age above 65 years are classified as older individuals and groups include more than three pedestrians moving to a similar direction. Pedestrians that carried a bag, held a child at their hand or towed a baby stroller moved more slowly (1.2 m/s on average) than pedestrians without these accessories. Maximum accelerations (4.5 m/s^2 on average) have been observed when pedestrians started to cross the street after they have been stationary waiting for the green traffic light signal or a sufficient gap between vehicles to cross. After having reached their natural walking speed, accelerations stayed below 1 m/s^2 on average.

There are differences for the speeds depending on the time of the day and the location. In the rush hour or around closing times of schools, pedestrians are forced to walk in groups and to interact. However, most of the individuals are in a hurry then and are focused on their destination, while pedestrians in the time in between tend to have more time and walk more slowly, e.g., since they are just on a shopping trip or on a promenade.

5.3.2 Stochastic Reachable Sets of Road Users

Section 5.2 presented a novel method to estimate the future path of the ego vehicle. The prior probabilities for different maneuver classes have been determined from statistics. Furthermore, the naive Bayesian classifier estimates a posterior probability for each maneuver class, although the probability might not be completely correct. However, it provides a good first indication. The predicted path from the presented approach and its probability are utilized now. Moreover, potential future paths for the two other maneuver classes are computed based on the speed-dependent prototype trajectories and the corresponding best start point from the LCS comparison. Thus, it is not required to know exactly which maneuver path the ego vehicle will follow, since several paths can be assumed with different probabilities at the same time. Opposite directions

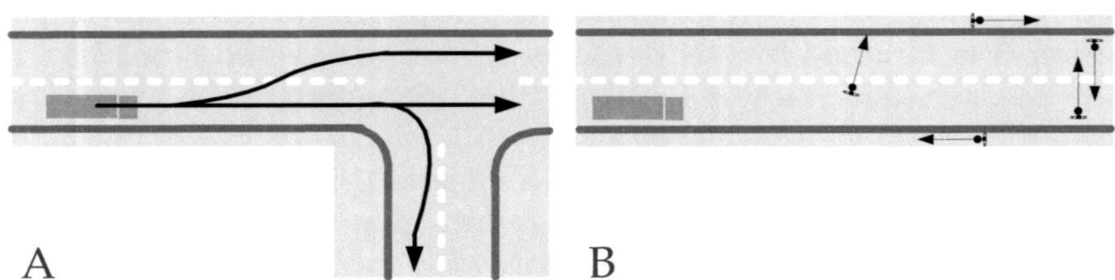

Figure 5.9 Examples for potential paths of the ego vehicle (A) and pedestrians (B).

are not taken into account, since left turn or left lane change maneuvers have never been mistaken for right turn or right lane change maneuvers and vice versa. An example for resulting possible future paths of the ego vehicle is shown in Figure 5.9 A.

If the truck driver follows one of the paths, he will probably not keep a constant speed. For example, he will decelerate before a turn maneuver and accelerate afterward. Moreover, he might want to pass a traffic light before it switches its lights from green to red and therefore accelerates. Furthermore, the ground or the vehicle itself may induce speed changes. Deviations from the predicted paths are obvious, since the truck never follows these perfectly. The prediction error — and thus the potential path deviation — increases with increasing distance from the current location. The lateral standard deviation from the actually driven path in meters depends on the maneuver class. The values depend on the prediction horizon and the maneuver class.

Pedestrians follow straight path segments more or less and keep their speed constant most of the time once they have reached their walking speed (see Section 5.3.1). However, there might be incidents like an approaching car or slower persons they know that lead to smooth acceleration or deceleration. Higher acceleration values are observable when a pedestrian starts to cross the street after stationary waiting at the roadside. A significant change in the direction could only be observed when pedestrians followed the road and tried to cross it then. Of course, slight path deviations are present due to the natural motion behavior of pedestrians or oncoming other pedestrians, and this is taken into account. The direction of the straight predicted pedestrian path is estimated as moving average from the pedestrian direction of the last measurement cycles. The current coordinate system has its origin in the middle of the rear axle of the truck, so that the relative pedestrian directions from the environment perception are stored in a ring buffer and are transformed to this coordinate system using the yaw rate signal and the speed signal of the ego vehicle. Thus, the uncertainty that results from the inaccuracy of these signals shall be considered in the approach as well. The probability that a pedestrian suddenly changes his direction is expected to be very small. However, if she does, the measurement updates will reveal this change and induce a correction. Examples for typical pedestrian paths are displayed in Figure 5.9 B.

As mentioned earlier, the measurement of the obstacle positions and speeds is not always accurate. Therefore, the measurement values are modeled by probability distributions with non-zero standard deviations. The inaccuracies of the state estimations decrease for the applied sensors radar and monocular camera the closer the objects get to the ego vehicle. The real existence of the obstacles is subject to uncertainties, which is expressed by the PoE. The current position of the ego vehicle defines the origin of the coordinate system, so that there is no inaccuracy in its current position. However, the measurement of its dynamic parameters like speed and yaw rate also shows inaccuracies that have to be modeled by a proper distribution. Means are required to handle uncertainty in state estimation and object existence as well as in state prediction.

The motion along the predicted path is assumed to be independent of the lateral dynamics. This is reasonable, since the task of path following is more or less independent of the task of keeping the velocity along the path or the distance to someone ahead. This assumption simplifies the probabilistic determination of the stochastic reachable sets of the road users, as the lateral and the longitudinal probability distribution can be computed independently in small dimensions.

A piecewise constant probability distribution $f(\delta)$ describes the lateral dynamics of the road users as displacement from the predicted path, see [196] for a statistical analysis of the lateral displacement of vehicles on a road. For pedestrians, the lateral displacement from the path has been obtained as part of the study conducted in this work. The probability distributions are normalized to the lane width for vehicles or to the assumed path width for pedestrians, so that they can be applied to lanes and paths of different widths. The deviation probability from the path is considered as constant in time and is assumed to be independent of the probability distribution along the path. The joint probability $f(s, \delta)$ results from the product of the obtained lateral probability distribution $f(\delta)$ and the longitudinal probability distribution $f(s)$ $(f(s, \delta) = f(s) \cdot f(\delta))$. The distributions $f(\delta)$ and $f(s)$ relate to the deviation δ of the road user's center from the path and its position s on the path. The combined probability distribution is described in a curved, path-aligned coordinate system as in [186].

The probability distribution along the path is obtained from a dynamic model with position s, velocity v and the absolute acceleration a. The acceleration corresponds to the input command u and is normalized to values within $[-1, 1]$ where -1 represents full deceleration and 1 maximum acceleration. The function $\rho(s)$ (in m) maps the path coordinate s to the curvature radius. For vehicles, the radius of the path determines the tangential acceleration a_t for a given velocity v and limits the normal acceleration a_n because the absolute value of the combined accelerations has to be smaller than the maximum, absolute acceleration a_{max}. The acceleration dynamics change at a switching velocity of v_{sw}. For speeds $v > v_{sw} \wedge u > 0$ the aerodynamic drag is considered. The differential equations for the longitudinal dynamics is chosen according to [186], where the second acceleration case is not used for the determination of pedestrian dynamics ($v_{sw} \to \infty$):

$$\dot{s} = v, \tag{5.18}$$

$$\dot{v} = \begin{cases} a_{max} u, & \text{if } 0 < v \leq v_{sw} \vee u \leq 0, \\ a_{max} \frac{v_{sw}}{v} u, & \text{if } v > v_{sw} \wedge u > 0, \\ 0, & \text{if } v \leq 0, \end{cases} \tag{5.19}$$

under the constraint that $|\mathbf{a}| \leq a_{max}$, where $|\mathbf{a}| = \sqrt{a_n^2 + a_t^2}$, $a_n = v^2/\rho(s)$, $a_t = \dot{v}$. Only limited absolute accelerations a_{max} are possible (Kamm's circle). The constants a_{max} and v_{sw} are chosen dependent on the specific properties of the different classes of road users. The differential equations result in a hybrid system and non-linear continuous dynamics. Moving backwards is not considered and a path to the opposite direction should be chosen instead. The hybrid automaton that represents pedestrian motion is visualized in Figure 5.10.

The stochastic reachable positions of other road users and of the ego vehicle within a certain time interval are computed to retrieve the probability of a crash as intersection of the stochastic reachable sets. Monte-Carlo simulations that utilize the presented dynamics can lead to quite accurate distributions of the predicted states. However, a higher number of simulation runs increases the accuracy, so that the computational effort may become giant. The requirement for real-time capability cannot

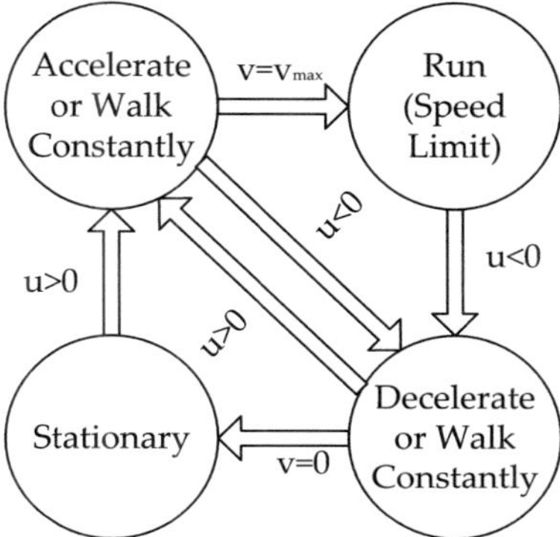

Figure 5.10   Hybrid automaton modeling pedestrian dynamics.

be satisfied if the accuracy is to be high. Therefore, Markov chains abstract the original dynamics with sufficient accuracy supporting the presented deterministic computations. A large number of simulation runs is performed during **offline computation** when the computation time is almost unlimited. Thereby one obtains a transition matrix for each class of road user with accurate state transition probabilities that can be saved in a database. The system contains only few dimensions (2) here. Thus, Markov chain abstraction is possible without extreme demand of storage and exploding computational complexity.

The continuous state space (position and speed along a path) and the input space are sampled to single cells. A unique map of the continuous to the discrete state of a hybrid system exists if the invariants of a hybrid system do not intersect, which is the case for the applied dynamics model. The bounded, discrete and two-dimensional state space is transformed to a vector where each element represents one state space cell for a simpler notation.

A bounded set of initial states is generated from a pre-defined grid on the initial cell $\mathcal{X}_j$. An input value $u([0,\tau])$ from a grid on the cell $\mathcal{U}^\alpha$ is generated for each initial state then. It is constant during the time interval $[0,\tau]$. Finally, every initial state is simulated with the inputs $u([0,\tau])$ and the parameter vector ρ of the time interval $[0,\tau]$ based on the system dynamics. The probability distribution within all state and input cells is strictly uniform. The assumption of a uniform distribution enables the reconstruction of a piecewise constant probability distribution

Parameter		Truck	Pedestrian	Car	Bicycle
v_{max}	in m/s	18	6	18	9
v_{sw}	in m/s	4	-	7.3	1
a_{max}	in m/s^2	7	5	9	7
s_{detect}	in m	70	70	70	70
t_{h}	in s	3.0	3.0	3.0	3.0
segment width	in m	0.5	0.5	0.5	0.5
segment length	in m	1	1	1	1

Table 5.1 Parameter definition: adapted from [71] for pedestrians based on sensor information and own study.

of the continuous state $\mathbf{x}$ from the discrete distribution of the Markov chain. The transition probabilities of the time step $\Psi_{ij}^{\alpha}(\tau)$ and of the time interval $\Psi_{ij}^{\alpha}([0, \tau])$ are computed as described in Section 3.6.

The input has to stay within the input cell U^{α} for the time interval $[t_k, t_{k+1}]$ ($t_k = k \cdot \tau$), but it may change for the next time interval. The Markov chain of each road user class is updated for the prediction horizon t_{h}.

The discretization region of the state space $\mathcal{X} = \bar{s} \times \bar{v}$ and the input space $\mathcal{U}$ shall represent the movement of all relevant road users for the whole temporal prediction horizon t_{h}. The total speed range reaches from standstill to the maximum considered speed $\bar{v} = [0, v_{\mathrm{max}}]$ and thus, the maximum position range is $\bar{s} = [0, v_{\mathrm{max}} \cdot t_{\mathrm{h}} + s_{\mathrm{detect}}]$, where s_{detect} is the maximum distance in which other road users (e.g., pedestrians) can be detected. The dynamics depend on a parameter vector ρ of the road user class from a set $\mathcal{P}$ and on the input u from the bounded set $\mathcal{U} \subset \mathbb{R}^m$. Table 5.1 visualizes the applied values.

When other road users are detected by the environment perception module during **online operation**, two transition matrices are loaded from a database for each road user dependent on its class (e.g., pedestrian). The normal distributions of the state estimation are transformed into piecewise, constant probability distributions in the discretized state space to obtain the initial probability vector $\mathbf{p}_0$. State cells outside of the 3σ-threshold (99.73 %) around the mean of the normal distribution get assigned a probability of 0.

The probability vector of the next time step $\mathbf{p}_{k+1}$ can be predicted by (compare Chapter 3.6.2)

$$\mathbf{p}_{k+1} = \mathbf{\Psi}^\alpha(\tau)\mathbf{p}_k. \tag{5.20}$$

It serves the prediction of the probability vector of the next time interval

$$\mathbf{P}_{[k,t_{k+1}]} = \mathbf{\Psi}^\alpha([0,\tau])\mathbf{p}_k. \tag{5.21}$$

The iterative multiplication of the probability distributions with the transition matrices evokes an error because the probability distribution within a cell is treated as if it was replaced by a uniform distribution in the next time step. The reduction of this error is possible by choosing smaller discretization samples and increasing the computational effort.

A time-dependent input transition matrix $\mathbf{\Gamma}_k$ can take into account uncertain input and thus, behavior models.

Pedestrians walking at normal speed (0.7 m/s to 1.7 m/s) are basically expected to keep their initial speed and the probability of acceleration or deceleration is small for the whole prediction horizon. Stationary pedestrians are expected to either keep standing or to accelerate fast, so that the probability for medium acceleration values is low (deceleration not possible). If the pedestrians accelerate, they probably keep walking at normal speeds once they have reached the normal range. Pedestrians with speeds up to 0.7 m/s are probably either in an acceleration phase or in a deceleration phase. Thus, the probability of keeping the speed constant is low. Faster pedestrians (faster than 1.7 m/s) have assigned a higher probability for deceleration than for acceleration until they reach the normal speed range.

Pedestrian interaction might lead to the fact that one pedestrian waits for another or the other accelerates to decrease the distance. Usually, pedestrians at the road edge that want to cross the street without infrastructural crossing decelerate to watch out for approaching cars if there are no other pedestrians on the street. A traffic light that switches from green to red might induce an acceleration of the pedestrians on the crossing, since they are running out of time. Of course, the detection of these events, e.g., by using car-to-infrastructure communication, is a prerequisite.

A normal speed range for turn maneuvers of the ego vehicle is 10 km/h to 15 km/h, as can be seen from the recorded data. Thus, a driver decelerates or accelerates to that speed range before the crossing, where acceleration is required if the driver had to stop at a crossing before turning, e.g., due to a red light. After having passed the crossing, he accelerates up to normal speed. Moreover, the recorded data showed that lane changes are usually driven at a constant speed, while lane following had the highest variability in speed and acceleration. The maximum speed is given by the predicted curvature profile $\rho(s)$ that could be based on GPS and map data if it was available. Applied values for the speed profile will be provided in Section 6.1.

The Monte-Carlo abstraction is not complete which is not required yet. Therefore, special algorithms that have been designed for sparse matrix multiplications like in [76] can accelerate the computation of the matrix products. Negligence of small probabilities in the transition matrix and the probability vector $\tilde{p}$ ($\tilde{p}_i < \epsilon_p$) and subsequent normalization to a sum of one results in these sparse probability matrices and vectors with only few non-zero entries. The threshold probability ϵ_p should be related to the combined number of input cells N_u and state cells N_x:

$$\epsilon_p = \frac{\epsilon_{p*}}{N_u \cdot N_x}. \tag{5.22}$$

Finally, one obtains the stochastic state distributions of all detected road users and the ego vehicle. However, since the abstraction is not complete, some occupied positions might appear empty. As a result, the risk could be under-estimated because the collision risk is computed by intersection of the stochastic position distributions of the ego vehicle and of the de-tected road users. Therefore, the set of longitudinally reachable positions is efficiently computed online. The initial state set is represented by a two-dimensional interval. Two simulation runs are sufficient to obtain the reachable positions given the presented dynamics, as the initial state jointly contains the maximum initial position and speed. If two sets of reachable positions intersect but the corresponding stochastic reachable sets do not due to incomplete Markov chain abstraction, one assumes a small value for the crash probability (compare [71]).

A vehicle with the initial condition $\overline{x}(0) = \mathcal{X}^0 = \overline{s}(0) \times \overline{v}(0)$ drives along a straight path following the dynamics $f(x(\tau), u(\tau)) = \dot{x}$ ($x \in \mathbb{R}^2$) of equations (5.18) and (5.19). $\overline{s}(0) = [\underline{s}(0), \overline{s}(0)]$ is the initial position interval and $\overline{v}(0) = [\underline{v}(0), \overline{v}(0)]$ is the initial speed interval. The reachable two-dimensional interval of the vehicle state $\overline{x}(t) = [\underline{x}(t), \overline{x}(t)]$ for time point t is given by

$$\underline{x}(t) \;=\; \underline{x}(0) + \int_0^t f(x(\tau), u(\tau))\mathrm{d}\tau, \; u(\tau) = -1, \tag{5.23}$$

$$\overline{x}(t) \;=\; \overline{x}(0) + \int_0^t f(x(\tau), u(\tau))\mathrm{d}\tau, \; u(\tau) = 1, \tag{5.24}$$

where $u \in [-1, 1]$ is a Lipschitz continuous input. Although the resulting state interval $\overline{x}(t)$ is an over-approximation of the exact reachable set of the state, one obtains the exact reachable set for the position. In case of a curved path, the tire friction is considered by determination of the minimum and maximum admissible input values for a given curvature profile $\rho(s)$, compare [197]. The replacement of $u(\tau) = -1$ by $u(\tau) = \underline{u}(s(\tau))$ and $u(\tau) = 1$ by $u(\tau) = \overline{u}(s(\tau))$ enables the identification of the reachable position and speed on a curved path. Cutting off the previously computed speed profile $v(s)$ at $v_{\max}$ accounts for the consideration of the speed limits ($\overline{u}(s) = 0$, if $v(s) > v_{\max}$). The maximum velocity is set to 65 km/h for trucks and passenger cars, because it is expected that the drivers more or less respect the speed limit of urban areas. Thus, the longitudinally reachable position interval $\overline{s}$ and speed interval $\overline{v}$ can be computed analytically for a given input interval $\overline{u}$ that is constant for the period of one computational time step t.

The analytical solutions of the longitudinal dynamics of the road users for $u \leq 0$ or $0 < v \leq v_{\mathrm{sw}}$ is given by

$$s(t) \;=\; s(0) + v(0)t + \frac{1}{2}a_{\max}ut^2, \tag{5.25}$$

$$v(t) \;=\; v(0) + a_{\max}ut \tag{5.26}$$

and

$$s(t) \;=\; s(0) + \frac{(v(0)^2 + 2v_{\mathrm{sw}}ut)^{\frac{3}{2}} - v(0)^3}{3v_{\mathrm{sw}}u}, \tag{5.27}$$

$$v(t) \;=\; \sqrt{v(0)^2 + 2v_{\mathrm{sw}}ut} \tag{5.28}$$

represents the analytical solution for $u > 0$ and $v > v_{\mathrm{sw}}$. The analytical solution for the case $v = 0$ is trivial.

The computationally expensive part can be accomplished offline when the computation time is almost unlimited, while the probability distribution can be computed efficiently by sparse matrix multiplications during online operation in the vehicle.

5.3.3 Conditional Collision Probability

The intended purpose of the procedure is to identify the collision risk of the ego vehicle with a vulnerable road user within the prediction horizon t_h if the driver does not suddenly change his driving style. If new sensor data are available after each time interval Δt, the update of the collision probabilities according to the new sensor information has to be performed for the whole prediction horizon t_h in that time as well. First own results for the determination of conditional crash probabilities have already been presented at scientific conferences [209,210]. The crash probability is a conditional probability in the sense of assuming that no previous collisions took place. The variables of the ego vehicle are indexed with $\Box_e$ for a better distinction between the variables of the ego vehicle and other road users $\Box_{ru}$. The collision probability with one road user is considered as a partial probability. The sum of the partial probabilities of all detected road users builds the total conditional collision probability.

Let $f(\boldsymbol{\xi}, t_k)$ be the probability distribution of a road user at time point t_k, where $\boldsymbol{\xi}_{ru}$ denotes a two-dimensional position vector of the road user center. As in [71], let $\mathrm{ind}(\boldsymbol{\xi}_{ru}, \boldsymbol{\xi}_e)$ be the indicator function that is 1 if the body of the ego vehicle and the body of another road user intersect and 0 otherwise. Then, then conditional crash probability $\tilde{p}_k^{\mathrm{crash}}$ can be formulated by

$$\tilde{p}_k^{\mathrm{crash}} = \int_{\mathbb{R}^\eta} \int_{\mathbb{R}^\eta} f(\boldsymbol{\xi}_{ru}, t_k) \cdot f(\boldsymbol{\xi}_e, t_k) \cdot \mathrm{ind}(\boldsymbol{\xi}_{ru}, \boldsymbol{\xi}_e) \mathrm{d}\boldsymbol{\xi}_{ru} \mathrm{d}\boldsymbol{\xi}_e. \tag{5.29}$$

The computational realization for the computation is described in the following. The probability distributions of the road user $f(\boldsymbol{\xi}_{ru}, t_k)$ and the ego vehicle $f(\boldsymbol{\xi}_e, t_k)$ at time point t_k are piecewise constant probability distributions in $\mathbb{R}^2$, since the path and the deviation from the path have been segmented and the distributions are uniform within one segment. Each path is simplified by connected straight lines $\overline{s}_g$ based on the extraction of points in appropriate path segment distances, where g is the path segment index. The deviation from the path is also subdivided into segments $\overline{d}_h$ with deviation index h. $\mathcal{C}_{gh}$ denotes the trapezoidal region that is spanned when the path coordinate s of the road user's center is within $\overline{s}_g$ and the corresponding deviation coordinate δ is within $\overline{d}_h$. The region that is occupied by the road users's body then is represented by $\mathcal{B}_{gh}$. The probability p_{gh}^{pos} that the center of a road user is within $\mathcal{C}_{gh}$ depends on the product

$$p_{gh}^{\mathrm{pos}} = p_g^{\mathrm{path}} \cdot p_h^{\mathrm{dev}}, \tag{5.30}$$

due to the independence assumption. Here, $p_h^{\text{dev}} = P(\delta \in \overline{\underline{d}}_h)$ represents the probability of the deviation segment and $p_g^{\text{path}} = P(s \in \overline{\underline{s}}_g)$ the probability of the path segment. The probabilities of the path segments p_g^{path} are retrieved from the joint probability distributions p_i^α of the state and the input resulting from the Markov chain computations. Summing up over all inputs provides the probability $p_i = \sum_\alpha p_i^\alpha$ of state space cell i, where each state space cell i represents a position interval $\overline{\underline{s}}_g$ and speed interval $\overline{\underline{v}}_l$ ($\overline{\underline{\mathbf{x}}}_i = \overline{\underline{s}}_g \times \overline{\underline{v}}_l$).

However, one is only interested in the probability of the path segments $\overline{\underline{s}}_g$ for the computation of the collision risk. Therefore, one integrates over all speeds to obtain the probability of a certain position on the path $p_g^{\text{path}} = \sum_l P(s \in \overline{\underline{s}}_g, v \in \overline{\underline{v}}_l)$. Now, the probability that the center of a road user is within $\mathcal{C}_{\text{ru},gh}$ is provided. However, the probability p_{efgh}^{int} that the bodies of the ego vehicle and another road user intersect is required as well. Therefore, the uncertain sets of the ego vehicle $\mathcal{C}_{e,ef}$ and the other road user $\mathcal{C}_{\text{ru},gh}$ are gridded uniformly. The gridding points represent potential centers of the road users where the bodies are located symmetrically around. The probability p_{efgh}^{int} is computed by counting the relative number for which the bodies of the road users would intersect. Finally, the partial, conditional collision probability can be computed by

$$\tilde{p}^{\text{crash}} = \sum_{e,f,g,h} p_{efgh}^{\text{int}} \cdot p_{\text{ru},gh}^{\text{pos}} \cdot p_{e,ef}^{\text{pos}}. \tag{5.31}$$

The sum is taken over all possible combinations of e, f, g, h resulting in a giant number of possible combinations. The computation can be accelerated if the two-step approach proposed by [71] is applied, since only a subset of index combinations has to be considered then. First, the approach checks for the intersection of road user bodies $\bigcup_h \mathcal{B}_{\text{ru},gh}$ belonging to a path segment $\overline{\underline{s}}_g$ and bodies of the ego vehicle $\bigcup_f \mathcal{B}_{e,ef}$ belonging to path segment $\overline{\underline{s}}_e$ by checking for the intersection of circles that enclose the corresponding set of vehicle bodies. Then, paired sets of vehicle bodies $\mathcal{B}_{\text{ru},gh}$ and $\mathcal{B}_{e,ef}$ that passed the first test are checked for intersection again by using enclosing circles. Look-up tables for the intersection probabilities p_{ghef}^{int} that depend on the road user class, the relative orientation and the translation of uncertain centers $\mathcal{C}$ can accelerate the computation of the crash probability.

Now, the conditional collision probability of the ego vehicle with a pedestrian that will follow path j when the ego vehicle will follow path

q can be calculated under consideration of the pedestrian's PoE $p_{\text{ped}}^{\text{exist}} = p(\exists x)$:

$$p_{\text{ped},j,q}^{\text{crash}} = p_{\text{ped}}^{\text{exist}} \cdot p_{\text{ped},j}^{\text{traj}} \cdot p_{\text{e},q}^{\text{traj}} \cdot \tilde{p}^{\text{crash}}, \tag{5.32}$$

where $p_{\text{ped},j}^{\text{traj}}$ represents the probability that the pedestrian will follow path j and $p_{\text{e},q}^{\text{traj}}$ denotes the probability that the ego vehicle will follow path q, e.g., a straight path.

Thus, the total conditional crash probability for one pedestrian in the considered time interval is given by numerical integration over all potential pedestrian paths and ego vehicle paths:

$$p_{\text{ped}}^{\text{crash}} = \sum_q \sum_j p_{\text{ped},j,q}^{\text{crash}}. \tag{5.33}$$

Here, the total conditional crash probability has been described exemplary for a pedestrian, but it works analogously for other road users. Clearly, the computation of a crash probability is only possible if the surrounding objects are not occluded and are perceived early enough. The objects have to be located within the FOV of the perceiving sensors. On the one hand, computing the partial conditional collision probability based on the probability distributions of the road users within consecutive time intervals implicates the advantage that no point of time is missed. On the other hand, the uncertainties are higher than for time point solutions, what may induce wrong collision probabilities. The presented approach is applicable to time step and time interval solutions.

The discretization of the state space and the input space for the presented Markov chain approach introduces a systematic error leading to an over-estimation of the stochastic reachable sets. Thereby the values of the crash probabilities can be under-estimated, since the relative intersection of the ego vehicle's reachable set and the one of another road user decreases, especially, in crossing scenarios (see Section 6.3). This impact can be reduced by choosing very small discretization intervals, which increases the computational effort. However, digital signal processors can accelerate the computations significantly, since the Markov chain approach is based on many matrix multiplications. The error in the transition probabilities can be chosen arbitrarily small, as the transition probabilities are computed beforehand and offline when computational time and resources are almost unlimited.

6 Results

This chapter is dedicated to the presentation of quantitative and qualitative results of the developed filters EKF-JIPDA and EKF-GNN for sensor data fusion and pedestrian tracking. Furthermore, the novel approaches for trajectory prediction and risk assessment are evaluated using exemplary scenarios and are compared to established approaches, such as trajectory prediction based on the assumption of a constant yaw rate and the computation of the TTC as risk measure.

6.1 Object Tracking and Object Classification

The perception performance of the proposed fusion approaches and implementations is evaluated to compare their strengths and weaknesses. The accuracy of the state estimation and the detection performance determine the perception quality.

The ideal evaluation of the object tracking procedure would require the exact state vector of each object, which is not available for real-world scenarios. A reference system has to be perfectly calibrated to the system that should be evaluated. Each sensor system that provides reference information for the state vector is subject to uncertainty as well.

Therefore, simulated sensor data is commonly applied to evaluate different filter approaches where the real state vector is exactly known. However, the evaluation results are not always comparable to the performance of the evaluated filters in real world scenarios with real sensor data. Simulation data can only model the simulated effects but not the real world effects that might sporadically appear. Noise parameters of the sensors and the total process cannot be represented in sufficient accuracy. False alarm rates and detection rates of sensors as well as their dependencies are not reproduced by simulation in a realistic way.

Therefore, an evaluation approach using real data has been chosen in this work. Various complex scenarios with pedestrians have been recorded in an urban environment and on a test ground. Tests at different vehicle speeds with pedestrians following or crossing the lane have been run. Variations of pedestrian speed and orientation have been evaluated as well as occlusion of pedestrians by other objects. Furthermore, the tracking performance has been evaluated in situations with multiple pedestrians and pedestrian groups in a scene. Locations with an increased probability for false alarms of the radar (e.g., ground with gravel) have been chosen as well as different illumination conditions influencing the detection performance of the camera. Different surface types and varying acceleration behavior of the ego vehicle induce various dynamic pitch angles that have an impact on the accuracy of the state estimation, especially, on the accuracy of the camera's position estimation.

Image data (>12000 samples) of these situations has been labeled manually to evaluate the detection performance of the two presented filter approaches — the EKF-GNN and the EKF-JIPDA. For the evaluation of the state accuracy of the pedestrians, a differential GPS system would have been beneficial. However, since no such system was available, the laser scanner was used as reference sensor for the positions. Measurements with pedestrians of known size enable the evaluation of the height estimation. A reliable reference for the pedestrians' speed was not available. Therefore, a qualitative evaluation of the corresponding state components will be presented here. The pedestrian speeds shall lie within a reasonable range — stationary pedestrians should be provided as stationary, slowly moving pedestrians shall have speeds around 1 m/s and below. A speed of about 1.5 m/s is considered as normal walking speed, while fast walking or running pedestrians should move with 2 m/s and above. For instance, pedestrian speeds should not exceed 4 m/s in these scenarios. Furthermore, the resulting motion directions should be plausible.

Since this work is not focused on the development of algorithms for the detection of pedestrians by camera or radar, the sensors' perception performance is taken as given, although there is still lots of room for improvement.

Both tracking approaches use the same state and measurement models and assume the same covariances for the sensors' measurements. However, evaluations showed that different parametrization is required for the EKF-JIPDA and the EKF-GNN regarding the threshold value of the HOG inference probability and the assumed uncertainty of the state model to obtain the best performance of each approach. The best performance means that the false alarm rate is kept below an admissible value. Therefore, the data has been filtered using the EKF-GNN with two parameter sets. The confidence values of the HOG detections may take on values between −1 and 1. For consideration of a HOG detection in the filter, the detection's confidence value had to exceed a certain threshold.

This threshold was set to 0.1 in the first parametrization of the EKF-GNN and to 0.7 in the second. The resulting filter variants are referred to as EKF01 and EKF07 in the following. HOG detections were included in innovation steps of the EKF-JIPDA if their confidence value exceeded 0.1. The abbreviation of this filter variant is referred to as JIPDA in the following.

Moreover, the JIPDA is parametrized with a larger uncertainty of the state model than the EKF01 or EKF07, which enables tracking in highly dynamic scenarios. However, the uncertainty of the state model has to be kept low for the EKF01 and the EKF07. Otherwise, tracks get assigned measurements from sources that are located at relatively large Euclidean distances if there was no association for some cycles, since the values of the state covariance matrix increase and thereby lower the Mahalanobis distance. These measurements are usually false alarms provided by the radar.[1] Consequently, the false alarm rate increases if the uncertainty in the state model is chosen too high. On the other hand, the uncertainty in the state model has to be kept at least as high that tracking across the sensory blind region is possible when objects change the direction of motion around this region. Furthermore, more missing detections have to be allowed around the sensory blind region than in the sensors' FOVs.

The perceptional power of the different filter approaches is presented and compared. The following subsection describes the procedure for evaluation of the state estimation. Subsequently, the parameters that provide evidence for the detection performance of the implemented approaches are introduced, while exemplary tracking scenarios are presented in Subsection 6.1.3 to compare the filter approaches in a qualitative way. Quantitative results of the evaluation of the state estimation and the detection performance are presented in Subsection 6.1.4 and in Subsection 6.1.5, respectively. The subsection concludes with the discussion of the results.

6.1.1 Methods for the Evaluation of the State Estimation

A common mean to evaluate the quality of a state estimation is the test for consistency. Consistency requires that the outcome of the procedure identifies the underlying truth [198]. Heuristically spoken, an estimator

---

[1] The JIPDA takes into account the detection probability of an object and the Mahalanobis distance in the PoE. Thereby objects are pulled to a smaller extent by distant measurements and are deleted earlier when only distant measurements are available than in the EKF01 or the EKF07.

is called consistent if an enlarged sample size brings the estimator closer to the real value of the estimated parameters. A state estimator of a dynamic system is consistent if the mean error of all points in time is 0 and the covariance of the state estimation equals the covariance of the estimation error. Mählisch [60] and Munz [65] tested their implementations of the JIPDA approach for consistency using an NCPS test (Normalized Projected Correction Squared) and an NEES test (Normalized Estimation Error Squared), respectively. The first is an adaption of the NIS test (Normalized Innovation Squared) to filters with weighted innovations. In contrast to the NEES test, the NIS and NCPS test do not require any reference data, but they require independent test samples, which is not given when using real measurement data. However, the statistics of the NIS or NCPS values can be tested by accumulation of sample values of different time points (histograms).

The NEES test assumes that the actual state vector $\mathbf{x}_{\mathrm{ref},ik}$ of object x_i at time point k (ground truth) is known. The NEES test checks whether or not the squared estimation error normalized to the state covariance matrix $\mathbf{P}_{ik}$ follows a χ^2 distribution with dim $(\mathbf{x})$ degrees of freedom. The NEES values $\epsilon_{\mathrm{NEES},ik}$ of object x_i at time point k can be computed using

$$\epsilon_{\mathrm{NEES},ik} = \left(\hat{\mathbf{x}}_{ik} - \mathbf{x}_{\mathrm{ref},ik} \right)^{\mathrm{T}} \cdot \mathbf{P}_{ik}^{-1} \cdot \left(\hat{\mathbf{x}}_{ik} - \mathbf{x}_{\mathrm{ref},ik} \right) \tag{6.1}$$

$$= \left(\boldsymbol{\gamma}_{\mathrm{ref},ik} \right)^{\mathrm{T}} \cdot \mathbf{P}_{ik}^{-1} \cdot \left(\boldsymbol{\gamma}_{\mathrm{ref},ik} \right), \tag{6.2}$$

where

$$\mathrm{E}\{\epsilon_{\mathrm{NEES},ik}\} = \dim (\mathbf{x}_{ik}) \tag{6.3}$$

should be fulfilled. Note that the estimation errors $\boldsymbol{\gamma}_{\mathrm{ref},ik}$ may be temporally correlated. A temporal computation of the NEES values gives evidence for filter consistency in sense of the NEES. The mean over all objects and all time points (N_{ref} samples)

$$\bar{\epsilon}_{\mathrm{NEES}} = \frac{1}{N_{\mathrm{ref}}} \sum_{n=1}^{N_{\mathrm{ref}}} \epsilon_{\mathrm{NEES},n} \tag{6.4}$$

approximates the expectation value for a large number N_{ref} and is therefore another indicator for filter consistency.

Here, an NEES test is applied to evaluate the filters' consistency. NEES values are computed using the state components with available reference data — longitudinal position, lateral position and pedestrian height. An NEES histogram results for each filter. The histograms are then compared to the density of a χ^2 distribution with three degrees of freedom. The comparison indicates whether the filter tends to be too pessimistic or too optimistic (tends to over-estimate its capabilities).

The consistency of a filter is influenced by several factors, such as imperfections of the process model or of the measurement model. Furthermore, numerical problems or mistakes in the implementation can induce filter inconsistency. The real measurement errors and process noise might not be normally distributed, or the model linearization can evoke a deformation of the distribution. If the parameters for the uncertainty of the state model (process noise) or the measurement noise do not exactly fit the theoretical values, filter consistency cannot be shown. Usually, the process noise has to be chosen pessimistically (over-estimated process noise) to enable tracking in highly dynamic scenarios. Thus, the distribution of the NEES values that can be obtained from the relative frequencies will have smaller values than the expected χ^2 distribution.[2] The obtained state uncertainty must not be estimated too small in a consistent filter.

The filter's underlying estimation error is an important statistical parameter for analysis of the filter's accuracy. The root mean square errors (RMSE) of the single state components l are computed to evaluate the dimensional estimation accuracy of the implemented filter approaches

$$\mathrm{RMSE}_l = \sqrt{\frac{1}{N_{\mathrm{ref}}} \sum_{n=1}^{N_{\mathrm{ref}}} \gamma_{\mathrm{ref},n}(l)^2}, \tag{6.5}$$

whereas the multi-dimensional estimation error provides a scalar quality measure of the state estimation

$$\mathrm{RMSE} = \sqrt{\frac{1}{N_{\mathrm{ref}}} \sum_{n=1}^{N_{\mathrm{ref}}} \|\gamma_{\mathrm{ref},n}\|^2}. \tag{6.6}$$

6.1.2 Methods for the Evaluation of the Detection Performance

The detection performance is evaluated by using manually labeled reference data from different scenarios. Thus, the real existence of each object

---

[2] Multi-model filters can reduce the effect and their development could be content of future work.

is known. All pedestrians in the image are labeled with boxes taking image area A_{label}, so that the position of the box in the image, its width and its height are known. Attributes are added to the labels such as for pedestrian occlusion by other obstacles, partial visibility at the edge of the FOV, or for pedestrians in a group. Each object hypothesis x_i is transformed from vehicle coordinates to image coordinates, where a constant pedestrian width of 0.4 m is assumed for each pedestrian. Each object projection can be represented by a box taking area A_{track} of the image. The overlap fraction o_{rel} of a labeled box and a box A_{track} resulting from an object hypothesis of the same time point is computed to determine if a pedestrian has been detected or if the object hypothesis is a false alarm. The overlap criteria is adopted from [65]:

$$o_{\text{rel}} = \frac{A_{\cap}}{A_{\text{track}} + A_{\text{label}} - A_{\cap}}. \tag{6.7}$$

An object hypothesis is considered as valid detection if o_{rel} exceeds the threshold value of 0.39 in this work. If a pedestrian object has been detected, it is a true positive detection (TP), whereas a non-detected object is classified as false negative (FN). If there has been an object hypothesis but no matching pedestrian, the object hypothesis is labeled as false positive (FP). Non-detected objects without object hypothesis would be true negatives (TN) but cannot be evaluated, since these situations are not explicitly detectable. All other obtainable numbers enable the computation of the true positive rate frac_{TP} and the false negative rate frac_{FN}:

$$\text{frac}_{\text{TP}} = \frac{N_{\text{TP}}}{N_{\text{TP}} + N_{\text{FN}}}, \tag{6.8}$$

$$\text{frac}_{\text{FN}} = \frac{N_{\text{FN}}}{N_{\text{TP}} + N_{\text{FN}}}. \tag{6.9}$$

Partially visible objects or partially occluded pedestrian objects are labeled as optional detections meaning that they may be detected (MB) but do not necessarily have to be detected. These are only counted in $N_{\text{TP}}^{\text{MB}}$ if there is a detection. Consequently, the true positive rate under consideration of all maybe-objects is given by

$$\text{frac}_{\text{TP}}^{\text{MB}} = \frac{N_{\text{TP}} + N_{\text{TP}}^{\text{MB}}}{N_{\text{TP}} + N_{\text{TP}}^{\text{MB}} + N_{\text{FN}}}. \tag{6.10}$$

An analogous computation of the false positive rate would require the availability of the number of true negative detections. Therefore, an alternative measure is used that normalizes the number of false positive detections to the number of iterations N_{it}.

$$\text{frac}_{\text{FP}} = \frac{N_{\text{FP}}}{N_{\text{it}}}. \tag{6.11}$$

The false positive rates of the tracking procedure should not be mistaken for the false alarm rate of an application. The object data is further pre-filtered before it can cause a system intervention. For instance, only pedestrians located around and within the driving corridor and fulfilling a certain speed criterion should be considered.

The computation of ROC curves enables the evaluation and interpretation of the filters' detection performance. If the PoE of a track has to reach a certain threshold ζ_{PoE} to be confirmed and only confirmed tracks are included in the number of true positive detections $(N_{\text{TP}} + N_{\text{TP}}^{\text{MB}})$, this has an impact on the true positive and false negative rates, see equations 6.10 and 6.11. If one varies the threshold ζ_{PoE} and plots the resulting detection rates of real pedestrians over the corresponding false positive rates, one obtains an ROC curve. Values for quantitative interpretation of the detection performance can be obtained by computing the AUC (area under curve), where higher values indicate a better detection performance than lower ones. Another measure would be the equal error rate. The value corresponds to the rate where the false positive rate and the detection rate are identical, so that the same unit or reference parameter is required for both rates.

Here, the AUC measure is utilized, since the true positive rates and the false positive rates relate to different reference parameters. The AUC corresponds to the area below the ROC curve up to a defined maximum false alarm rate, for instance 2 per iteration. Thus, the AUC of an optimal filter with a perfect ROC curve takes on the value 2 in this example. The AUC values of the filter variants JIPDA, EKF01, and EKF07 enable a quantitative comparison of the detection performance of the filters.

In the last two subsections, quantitative measures for the comparison of the EKF variants and the JIPDA have been presented. The following subsection introduces some example scenarios for a qualitative impression of the filters' performance. Quantitative results will follow in Subsection 6.1.4.

6.1.3 Examples for pedestrian tracking with EKF-JIPDA and EKF-GNN

The examples are presented as a combination of an image with projected JIPDA tracks (gray boxes) and the plots of the state components and the PoE in the Figures 6.1 to 6.6. The color of the box in an image represents the object's PoE. The lighter the box is, the higher is the PoE. Green boxes in the image represent HOG detections. Yellow circles around the foot points of the pedestrians illustrate the position covariances. The green horizontal line is the estimated horizon taking into account the low-frequent part of the dynamic pitch angle. In the plots, the thick solid blue lines represent object data of the JIPDA, while dashed gray and black lines show the tracking results of the EKF07 and the EKF01, respectively. Red circles indicate HOG detections. The sensory inference probability $p_{TP}(z_j)$ of the HOG detections is plotted together with the PoE of the tracks. Stars represent measurements from the radars, while magenta diamonds illustrate pedestrian detections of the laser scanner. Note that only detections of pedestrians on the right of the ego vehicle can be detected by the laser scanner if it is mounted on the right step tread.

The first exemplary scenario that is presented in Figure 6.1 has been recorded on a large tarred test ground, so that there were no other objects in the environment and only a very low number of false detections has been provided by the radars. The laser scanner has been mounted on the right upper step tread. Two pedestrians approach from the front on the left and on the right side of the truck. Then, they start to cross the driving corridor. The ego vehicle is running at about 15 km/h. The state estimation and detection performance of the JIPDA and the EKF01 is almost identical, since there are only a few radar reflections on the ground that could disturb the state estimation. The EKF07 deviates in the state estimation because it takes into account less HOG detections than the EKF01 and the JIPDA due to the HOG threshold. The longitudinal distance estimation is dominated by the radar, while the lateral distance follows the camera's HOG detections. The longitudinal speed of the objects takes longer to stabilize than the lateral speed but both appear plausible. The height estimation is acceptable, since the real pedestrians had heights of about 1.8 m and 1.9 m. The plot on the right bottom illustrates that the PoE's of the EKF07 and the EKF01 are pure heuristics, as described in Subsection 4.6.2. The PoEs of the JIPDA objects initially grow and stay at constantly high values in this scenario, which is plausible since there is no disturbance that could lead to ambiguities in this situation.

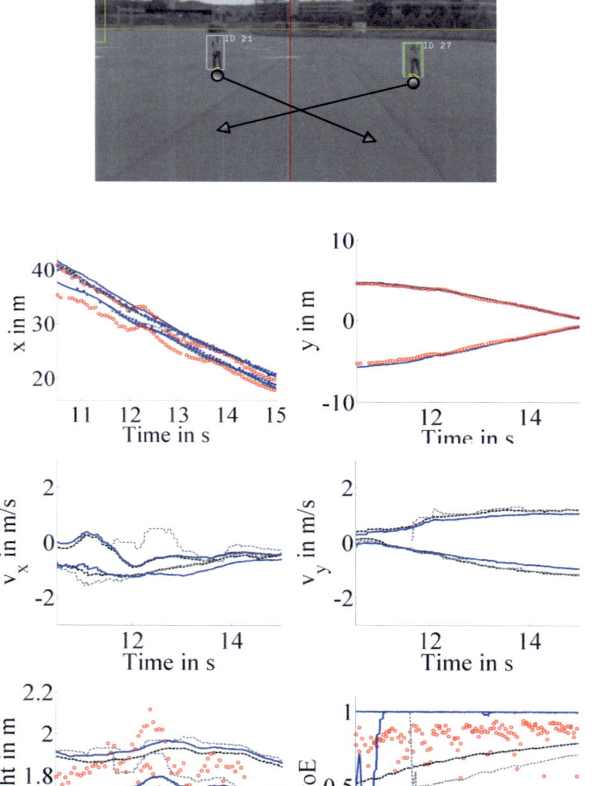

Figure 6.1 Scenario where two oncoming pedestrians follow the lane before they start cross it in front of the ego vehicle. The ego vehicle is running at about 15 km/h. The state estimation of the filter variants is similar, since there is no disturbance that could lead to ambiguities. (The legend is provided in the text at the beginning of Subsection 6.1.3.)

The second scenario includes three pedestrians. It is illustrated in Figure 6.2 where the plot on the right top shows the positions of two pedestrians in a world-fixed coordinate system. One pedestrian walks around the cottage in the background at the beginning of the scene. Two other pedestrians cross the lane in front of the ego vehicle on a ground with gravel. Radar reflections from the gravel and inaccurate distance information of the camera result in a lowered PoE of the closer pedestrian, so that the object is deleted and a new object is initialized. A false HOG detection on the vehicle in the background leads to a track with a low PoE that does not survive for a long time. This track has also been created by the EKF01, but no distinction between trustworthy and non-trustworthy is possible for the EKF01 due to the heuristic nature of the PoE.

The next scenario (Figure 6.3) has been recorded in a more crowded environment. Cars and elevated infrastructural elements on the street edge induce several radar detections and some HOG detections of non-pedestrian objects. The ego vehicle approaches a zebra crosswalk and decelerates to standstill. Two pedestrians cross the lane in front. There are two other pedestrians walking on the left pavement at a distance initially larger than 50 m. The EKF01 and the EKF07 induce more false positive objects or several objects that result from the same real pedestrian. This can happen when a HOG detection could not be associated with an existing track and a new track is set up for the pedestrian detection. If the spatial uncertainty of the first track is sufficiently large and there are other detections around — usually radar reflections — both tracks are assigned to measurements and survive. The plots do not show the radar detections. False positive objects of the JIPDA do not exceed a PoE value of 0.5, while true positive tracks showed PoEs above 0.5. Thus, the JIPDA is able to distinguish between false and true positives by using the PoE. This is not the case for the EKF-GNN filter variants, where all track types show more or less the same heuristic PoEs.

Figure 6.4 shows a group of three pedestrians that approaches on the left pavement. Initially all three have been tracked. Then, the height of the HOG detections deviates too much from the initial estimate of one object. Consequently, the object is deleted. The EKF-GNN tracks with the highest lateral speed represent false positive objects.

The scenario that is shown in Figure 6.5 and in Figure 6.6 illustrates the filter behavior for tracking across the sensory blind region and in the blind spot of the ego vehicle. One pedestrian approaches from the front and another crosses the lane from the left in front of the ego vehicle before he turns and enters the blind region. Sporadic false HOG detections on

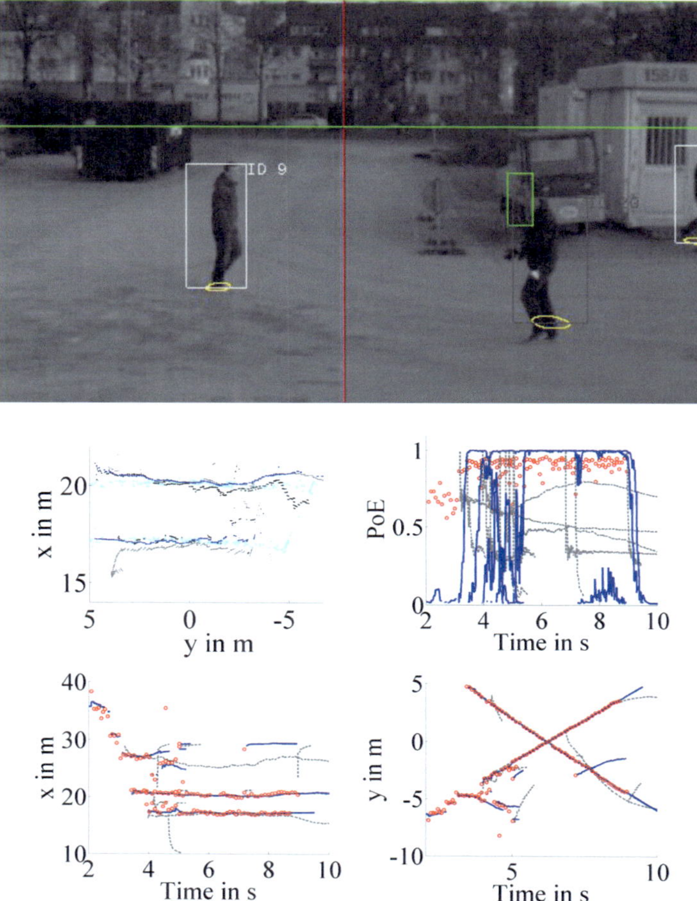

Figure 6.2 Scenario with two pedestrians crossing the lane and one pedestrian walking around the cottage in the background: The plots visualize the state components and the PoE over time of the JIPDA tracks (solid blue). The laser scanner has been mounted on the right step tread of the stationary ego vehicle. Cyan stars represent radar detections. Inaccurate HOG detections (red circles) lower the PoE of the closer pedestrian so that one object is deleted and another is initialized. The camera-based false alarm on the vehicle only induces an object with a low PoE.

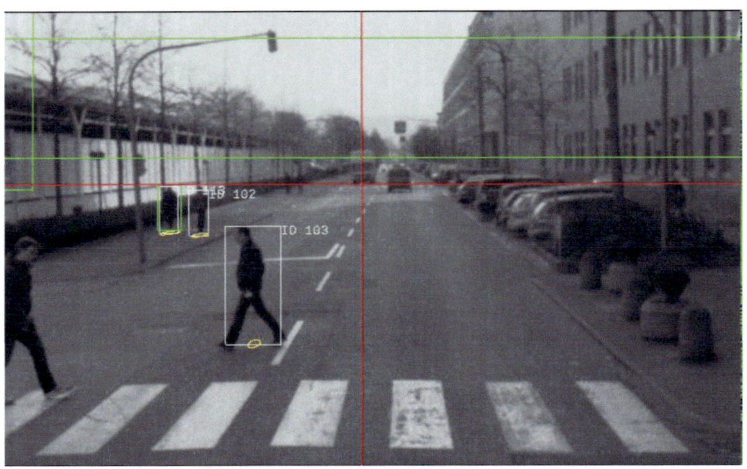

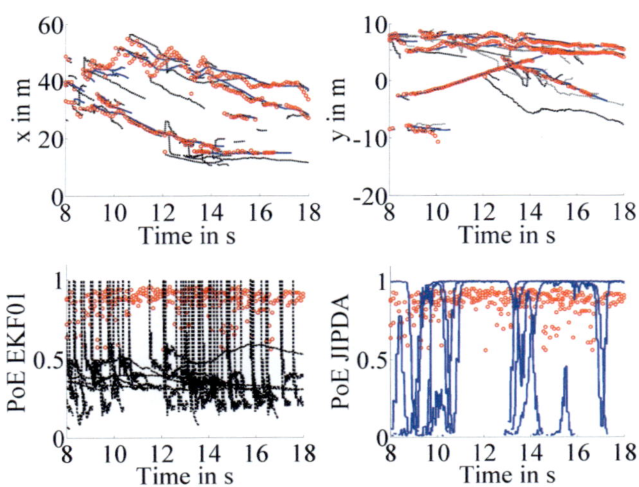

Figure 6.3 Scenario with four pedestrians, where two pedestrians cross the lane in front of the decelerating ego vehicle and two pedestrians walk on the left pavement at a relatively large distance (50-60 m). False positives of the JIPDA can be distinguished from true positives based on the PoE, which is not the case for EKF01 and EKF07 tracks. (The legend is provided in the text at the beginning of Subsection 6.1.3.)

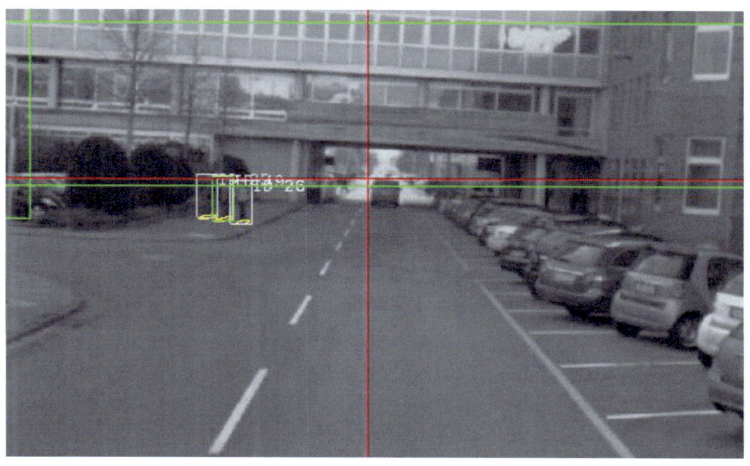

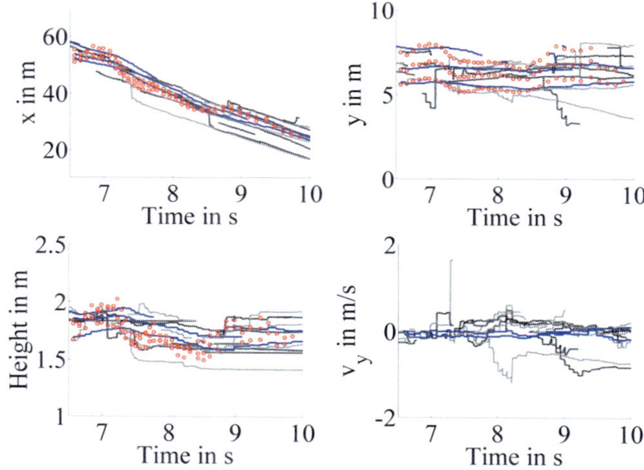

Figure 6.4 Scenario with a group of three pedestrians approaching on the left pavement. Two pedestrians could be tracked quite stable, while the track of the left pedestrian is lost after some time, since the height estimation of the camera strongly deviates from the initial height estimation. (The legend is provided in the text at the beginning of Subsection 6.1.3.)

the traffic sign induce the creation of tracks by all filter variants, but the corresponding JIPDA tracks never reach high PoEs and die after a few cycles, whereas radar reflections from that sign keep the EKF01 and EKF07 tracks alive.

When the pedestrians leave the field of detection of the camera there is no update for the height estimation of the pedestrians anymore. Thus, the value is kept constant. The decrease in the PoE of the JIPDA around second 20 results from several radar reflections from the rough ground with gravel around the pedestrian, which can be observed in Figure 6.6. These radar reflections also induce the effect that the tracks slightly deviate from the real object position, which can be seen in Figure 6.6 at the right edge of the SRR's FOV. The plot on the left shows the tracks of the JIPDA in world-fixed coordinates, while the plot on the right visualizes the corresponding tracks of the EKF07. Both filters are able to track the pedestrians across the sensory blind region. The tracks can be kept alive, so that the classification information of the frontal camera can be exploited in the blind spot of the vehicle and an increased confidence about the object class and the object height can be provided by the filters in the blind spot region.

Figure A.3 of Appendix A.8 shows images of additional scenarios using the JIPDA for tracking of multiple pedestrians and another exemplary scenario for the filter comparison where two pedestrians follow the lane on the right pavement. Figure 6.7 visualizes some challenging scenarios for object tracking. Only some pedestrians are detected by the camera and can be tracked if there are too many pedestrians in a scene. If pedestrians walk laterally in a group and the pedestrians partially occlude each other, only the closest pedestrian is detected by the camera. Pedestrians that are as close that the foot point is not anymore in the image are usually not detected. A pedestrian's height and distance cannot be estimated correctly by the camera if the lower part of a pedestrian is occluded or the pedestrian is located on a ground that is located higher than the ground plane of the ego vehicle.

The presented situations gave a qualitative impression of the filters' performance and indicated which challenges have to be handled. Quantitative results are presented in the following subsections.

6.1.4 Results of the Evaluation of the State Estimation

Scenarios with pedestrians of known height have been recorded, where the laser scanner has been mounted at the front for one set of scenarios

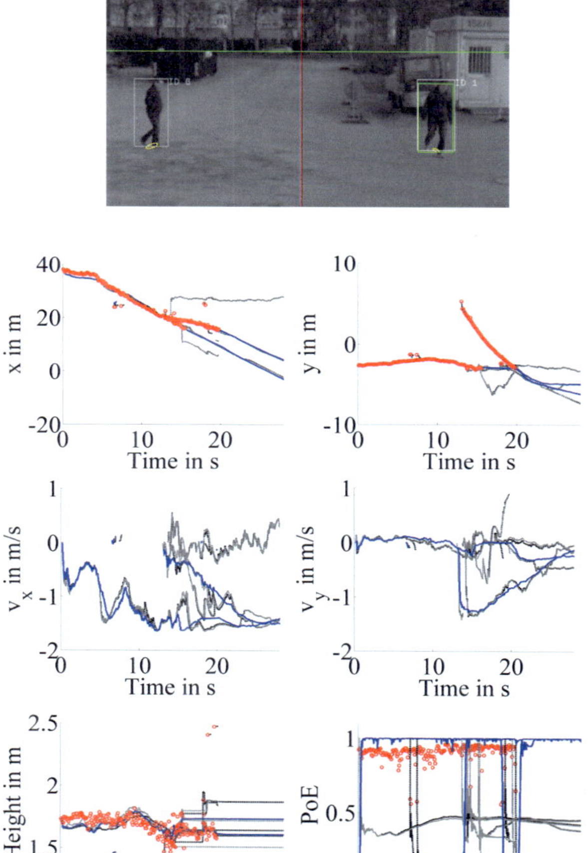

Figure 6.5 Scenario with one pedestrian approaching the blind spot from the front and one approaching from the left. The EKF variants stably track the traffic sign, whereas the JIPDA deletes the corresponding initialized tracks after a few cycles due to a low PoE. (The legend is provided in the text at the beginning of Subsection 6.1.3.)

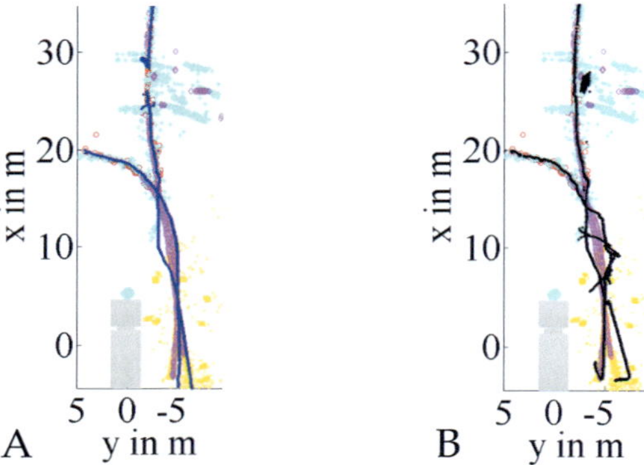

Figure 6.6 Pedestrian positions in a world-fixed coordinate system in a scenario, where one pedestrian approaches the vehicle's blind spot from the front and another from the left: Positions from JIPDA (A, solid blue line) and EKF07 (B, black line). The laser scanner detections are represented by magenta diamonds. The situation is the same as in Figure 6.5.

Figure 6.7 Examples for challenging tracking scenarios: Only some pedestrians are detected by the camera in crowded scenes. Pedestrian groups are tracked as one pedestrian if the single pedestrians occlude each other. The foot point of a pedestrian has to be within the image to enable a detection by the camera. Partial occlusion of pedestrians often hinders stable detection and tracking.

and on the upper right step tread in another set. The scenarios have been mainly recorded in an environment with few non-pedestrian objects to avoid false detections by the laser scanner, which could lead to an ambiguous association of detections with real pedestrians and other objects. Therefore, it has to be expected that the state estimation, especially of the EKF01 and the EKF07, is evaluated to be more accurate than it would be in more structured and crowded scenarios. All tracks that had a PoE > 0.01 have been considered for the evaluation. The pitch angle of the cabin has a recognizably negative impact on the perception performance of the laser scanner at large distances. Therefore, the evaluation is only performed for a limited distance range of 7 m to 38 m when the laser scanner is mounted at the front of the driver's cabin, although the filters track objects up to a range of 70 m.

Figure 6.8 shows the histograms of the reference positions used for the evaluation of the state estimation. The positions have been obtained from the pedestrian detections of the laser scanner mounted at the front. The real pedestrians used for data recording are 1.8 m and 1.9 m tall.

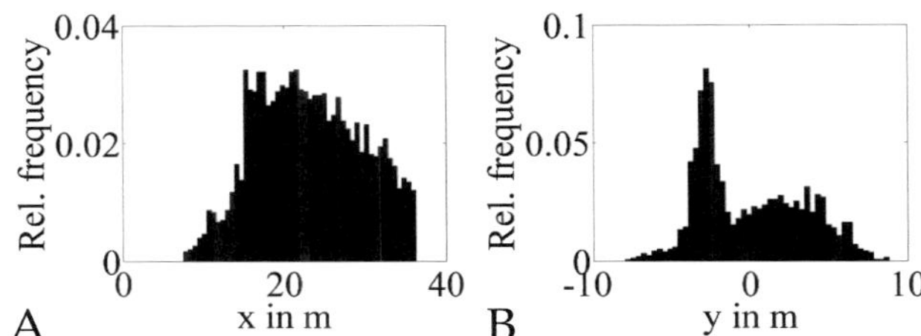

Figure 6.8 Histograms of the reference values for the positions (x, y) that have been obtained from the laser scanner.

The multi-dimensional RMSE has been computed for the different filter variants. Table 6.1 shows the results together with the RMSE computations of the single state components. The radars usually detect pedestrian legs, and the position of a pedestrian's foot point in the image depends on her step size. The laser scanner takes the center of mass of the computed cluster for the position, so that the reference data may deviate from the body center. Thus, the reference position of a pedestrian can depend on her current shape due to the current position of the arms and the legs. The obtained RMSE values appear as a good result for the utilized sensors, since the value for the longitudinal distance RMSE_x is around 0.5 m, which is less than a normal step size. The lateral value RMSE_y is even lower.

Filter	JIPDA	EKF07	EKF01
RMSE	0.54	0.59	0.52
RMSE_x in m	0.51	0.55	0.48
RMSE_y in m	0.14	0.18	0.15
RMSE_h in m	0.11	0.13	0.12

Table 6.1 Root mean square error analysis for the filter variants JIPDA, EKF07, and EKF01.

Figure 6.9 shows box-whisker plots[3] obtained from the deviations $\gamma_{\text{ref},ik} = \hat{\mathbf{x}}_{ik} - \mathbf{x}_{\text{ref},ik}$ between the estimated mean state and the corresponding reference data (laser scanner, known pedestrian height). The

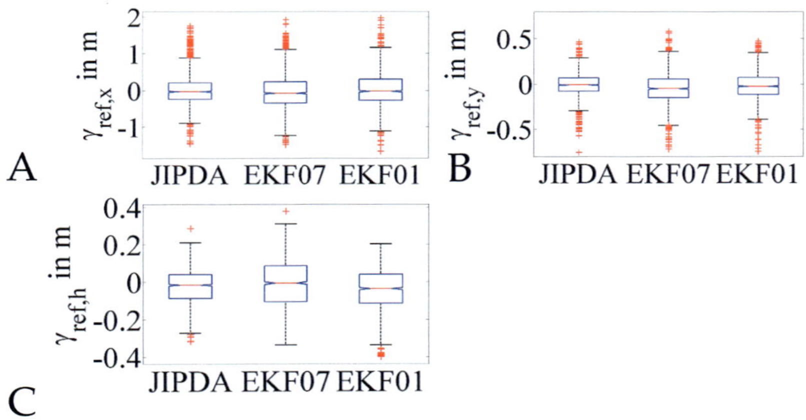

Figure 6.9 Box-whisker plots of the deviation $\gamma_{\text{ref},\square}$ of the estimated mean state from the reference data.

---

[3]Box-whisker plots visualize different statistic measures. The tops and bottoms of each box correspond to the 25th and 75th percentiles of the samples, respectively. The distances between the tops and bottoms are the inter-quartile ranges (IQR). The line in the middle of each box is the sample median. The whisker lines extend above and below each box. Observations beyond the whisker length are marked as outliers. An outlier is a value that is more than $1.5 \cdot \text{IQR}$ away from the top or bottom of the box.

corresponding values of the median as well as of the 25th and 75th percentile are summarized in Table 6.2. Notches of the box-whisker plots display the variability of the median between samples. The width of a notch is computed so that box plots whose notches do not overlap have different medians at the 5 % significance level. If one compares the medians of different box-whisker plots, one performs a visual hypothesis test. The deviations from the reference data stay in a closer range for the JIPDA

Filter		JIPDA	EKF07	EKF01
	25th percentile	-0.24	-0.35	-0.27
x in m	Median	-0.03	-0.08	-0.02
	75th percentile	0.21	0.24	0.30
	25th percentile	-0.08	-0.15	-0.11
y in m	Median	-0.01	-0.05	0.03
	75th percentile	0.07	0.05	0.07
	25th percentile	-0.09	-0.10	-0.12
h in m	Median	-0.02	-0.01	-0.04
	75th percentile	0.04	0.08	0.04

Table 6.2 Statistic parameters of the deviation of the estimated mean states from the reference data (laser scanner, ground truth).

than for the EKF01 and the EKF07. The EKF01 under-estimates the pedestrians' height significantly more than the JIPDA (p-value $7.0 \cdot 10^{-5}$)[4] or the EKF07 (p-value $3.3 \cdot 10^{-11}$), which is reasonable, since HOG detections that under-estimate a pedestrian's height usually have a smaller inference probability ($p_{\mathrm{TP}}(z_i) < 0.85$) than other HOG detections. The JIPDA explicitly considers this inference probability and the EKF07 only includes HOG detections above that value. On the other hand, since fewer detections are used by the EKF07 and the lateral distance accuracy is dominated by the camera, the lateral accuracy of the EKF07 is significantly lower than for the EKF01 (p-value $1.3 \cdot 10^{-7}$) and the JIPDA (p-value $4.8 \cdot 10^{-10}$), whereas the lateral accuracies of the JIPDA and the EKF01 do not differ significantly. The accuracy of the longitudinal distance estimation is comparable for the JIPDA and the EKF variants. However,

---

[4] Refers to Student's t test: The p-value is the probability of observing a value as extreme or more extreme of the test statistic $t = \sqrt{(m_a - m_b)/(\frac{s_a}{N} + \frac{s_b}{M})}$, assuming that the null hypothesis is true, where m_a and m_b are the sample means, s_a and s_b are the sample standard deviations, and N and M are the sample sizes.

Student's t test showed a significant difference between the EKF01 and the EKF07 (p-value 0.045).

As mentioned above, only a qualitative evaluation of the speed accuracy is possible due to missing reference data. Figure 6.10 illustrates the histogram of the speed values that have been obtained by the JIPDA. All speed values lie in a plausible range. Each histogram shows a maximum at 0 m/s, which can be explained by the fact that detected pedestrians were either standing or moved into the normal direction of the corresponding speed component. Local maximums can also be observed between 1 m/s and 2 m/s, which corresponds to the normal walking speed of a pedestrian. Single recordings included pedestrians jogging laterally or longitudinally. These pedestrians' speeds have been estimated to be about 3 m/s.

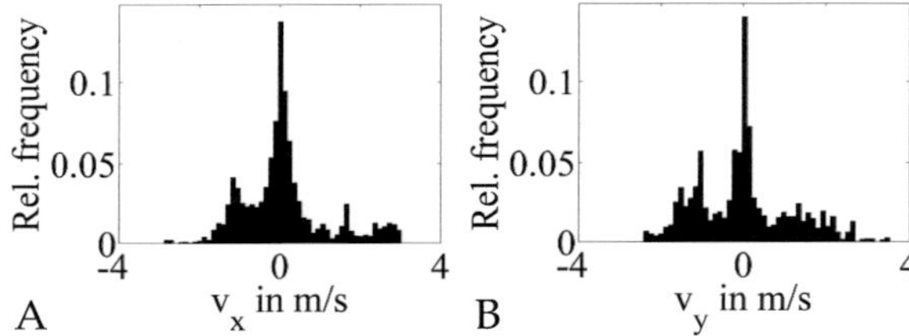

Figure 6.10 Histograms of speed values estimated by the JIPDA: longitudinal speed v_x (A) and lateral speed v_y (B).

The histogram of the computed NEES values is exemplarily shown for the JIPDA in Figure 6.11 together with the χ^2 density distribution with three degrees of freedom. The distribution indicates that the filter underestimates its accuracy and provides rather pessimistic values for the uncertainty of the state estimation. However, the limitations of the process model require the assumption of a high process noise to enable pedestrian tracking in highly dynamic scenarios. The filter never over-estimates its accuracy, so that the filter can be considered consistent. The EKF01 and the EKF07 show slightly higher NEES values but can be taken as consistent as well. Large deviations of the estimated mean state from the reference data come along with a high state uncertainty.

Summarizing, one can state that the accuracy of the state estimation of all filter variants is satisfying. All filters provide rather pessimistic estimations of the state uncertainty, but this is a required feature to enable pedestrian tracking in highly dynamic scenarios.

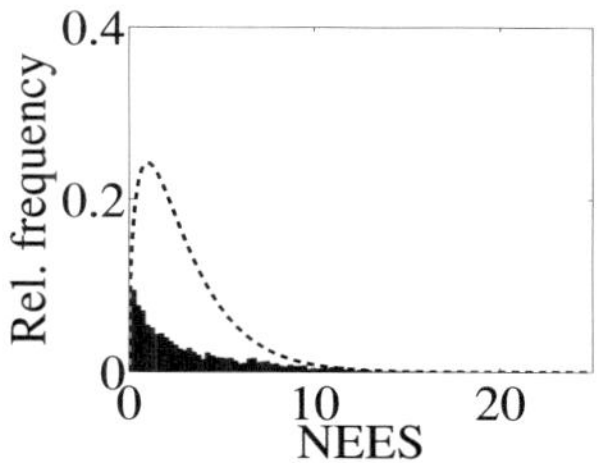

Figure 6.11 Histograms of the NEES values for the JIPDA and the χ^2 density distribution with three degrees of freedom (dashed line).

Another important aspect of the perceptional power of the filters is the detection performance, which is evaluated in the following subsection.

6.1.5 Results of the Evaluation of the Filter's Detection Performance

The manually labeled data described at the beginning of this section has been used for the evaluation of the detection performance. The object data from the filters has been interpolated to the timestamps of the label data (camera timestamps) to enable an association of the labels with the filter data as unambiguous as possible. Then, the ROC curve of each filter variant is computed as described in Subsection 6.1.2. Figure 6.12 shows the obtained ROC curves together with the corresponding evolution of the PoE. The PoE curves of the EKF01 and the EKF07 show that the chosen mean for the estimation of an object's confidence is only a heuristic, since both curves increase rather linearly. The detection performance of the EKF07 is superior to the detection performance of the EKF01, since it takes into account more reliable HOG detections. The plots indicate that only objects with PoEs > 0.5 should be provided to subsequent situation evaluation modules to keep the false alarm rate within an acceptable range. However, the detection rate is not very high then. A significantly better detection performance is obtained by the JIPDA, that shows a higher detection rate at a lower false positive rate.

Table 6.3 summarizes the scalar AUC values normalized to two false positives per iteration for the comparison of the detection performance of the different filter variants. The HOG detections show a better detection performance than the EKF01 and the EKF07, which can be explained by the high false positive rates of these filters. The high false positive rates result from object associations with radar targets resulting from non-

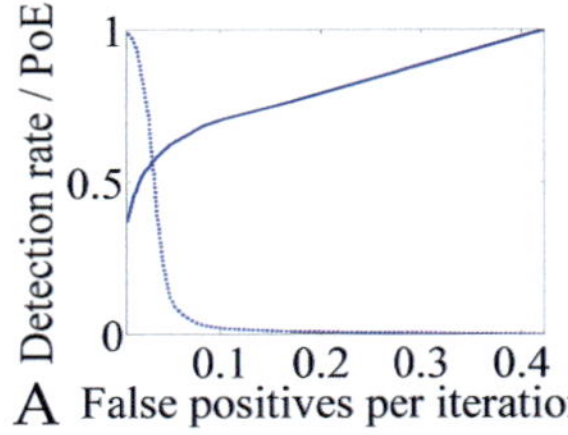
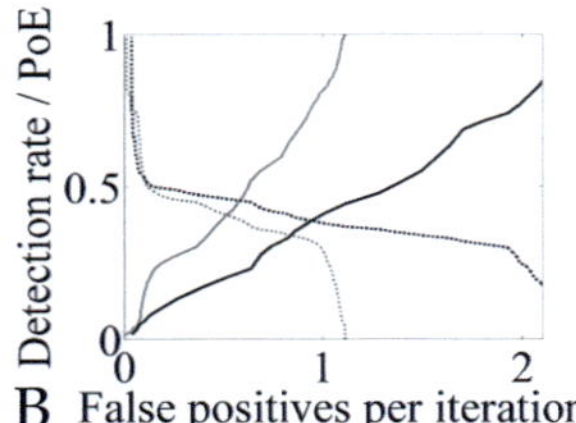

Figure 6.12 ROC curves of the JIPDA (A, blue solid line), the EKF07 (B, gray solid line), and the EKF01 (B, black solid line). The dashed lines show the corresponding evolution of the PoE.

pedestrian objects. This is less likely to happen in JIPDA filtering and therefore, this filter shows the best detection performance.

Filter Variant	JIPDA	EKF07	EKF01	HOG	Optimum
AUC	1.90	1.40	0.80	1.72	2.0

Table 6.3 Areas under curve normalized to 2 false alarms per iteration.

After a quantitative comparison of the filter approaches, the next sub-section summarizes the results, discusses the challenges that have to be faced, and provides ideas for improvements.

6.1.6 Conclusions from the Filter Comparison

All presented filter variants base their state estimation on the extended Kalman filter. The JIPDA, the EKF01 and the EKF07 are causal filters considering current data and data from the history. Initially, it takes some cycles to reach a sufficient confidence level (PoE) for track confirmation. During that time, a higher false negative rate is to be expected than for higher PoEs. After an object's disappearance it takes some cycles to decrease the PoE and to delete objects which can lead to an increased false positive rate. Therefore, the filters can be considered as smoothing filters.[5]

The accuracy of the state estimation is comparable for all presented filter variants, however, the JIPDA estimates are more accurate. The JIPDA

---

[5] The impact of different parameter sets for the probability of detection $p_D(x_i)$ and the sensory inference probability $p_{TP}(z_j)$ on the transfer behavior of the JIPDA filter has been analyzed in [60]. Therefore, it is not discussed in more detail here.

filter's detection performance is superior to that of the two other filter variants, which is an argument for probabilistic data association in multi-object tracking. The PoEs provided by the JIPDA filter appeared reasonable and decreased in situations where measurements from several clutter sources close to the object have been received. Objects that have been initialized by false detections only obtained low PoEs in JIPDA filtering. In contrast, the EKF01 and the EKF07 could only provide heuristic PoEs, so that one can only barely infer from it if the tracked object really exists. All filter approaches succeeded in tracking across the sensory blind region, although the position estimations of the JIPDA filter were located closer to the real pedestrian position than of the EKF01 and the EKF07. Thereby, information about an object's class and height could be provided for objects in the blind spot of the vehicle, information that would not be available if only one separate radar was used for this region. If the vehicle moved faster than 25 km/h and the pedestrians changed their direction significantly around the sensory blind region, the filters could not follow this behavior. However, since the vehicle will not turn or stop at this speed, the pedestrian objects in the vehicle's blind spot are not relevant anymore and thus, do not represent a risk.

So far, only the outcome of the filter approaches has been compared, but the preliminaries of the approaches and the effort should be discussed as well. The EKF-GNN variants are easy to implement and do not require a high number of parameters. The complexity of these algorithms is much lower than for the JIPDA.

Adding an additional sensor with another distinct FOV is possible in a generic way in JIPDA filtering and in the EKF-GNN approaches, but the utilization of the detection probability makes it more comfortable in JIPDA filtering. The detection probability that is 0 in the sensory blind region reduces the effect that tracks are pulled by measurements from sources that are located around but do not result from the object.

The JIPDA is much more sensitive to an incorrect calibration, as the detection rates have to fit to the FOVs. The Mahalanobis distance as well as the probability of detection have an impact on the PoE and thereby on the track management in JIPDA filtering. Furthermore, the local detection performance of a camera may depend on lens effects that cannot be compensated by intrinsic calibration and these effects may differ between different camera samples.

The calibration procedure of the sensors has to be kept simple to enable a calibration at the end of a production line and in workshops, for instance, if a windshield has to be changed. Usually, the rotation parameters should be computed only during online calibration then. The EKF-GNN approaches are more likely expected to be able to handle the demand for simplicity.

Camera and radar performance is taken as given. However, there is still room for improvement, for instance, in the radar sensors' design for pedestrian detection. The detection performance of the sensors strongly influences the total performance of the tracking approaches. The sensors have been measured under laboratory conditions at clear sight but the detection performance strongly variates with illumination conditions and in a crowded urban environment. For example, although the camera has an integrated intelligent shutter control algorithm, the detection performance for pedestrians strongly decreases during low altitude of the sun, since the prolonged shutter time increases the motion blur. Furthermore, the detection algorithm does not detect all pedestrians in scenes with a large number of pedestrians at once due to computational deficits. When several pedestrians are located close to each other, only one pedestrian of the group might be detected in one cycle and only one other pedestrian of the group in the next cycle, so that the initialization of stable tracks is unlikely. If the detection probability is lowered too much, detections in less crowded scenes are categorized more likely as false alarms. One has to find a compromise to obtain acceptable results for both environments. If there are multiple strongly reflecting objects at the roadside, such as parked cars or little walls, all available 64 radar detections might be used for representation of these objects and no target is set for a lowly reflecting pedestrian. Thus, the probability of detection would have to be modeled dependent on several other conditions that might even not be determinable with available sensors.

The image-based detection performance can be enhanced by extending the training data set of the HOG classifier by images from more different environments and illumination conditions. Moreover, a more sensitive camera with a shortened illumination time would decrease motion blur. Radar sensors with a higher sensitivity and an increased angular resolution (higher bandwidth) as well as extended computational capacities are under current supplier development. The expectations assume that the detection performance will be highly increased and that the application of micro-Doppler[6] approaches will even enable radar-based pedestrian

---

[6] Relative motion or deformations of parts of illuminated objects induce additional features in the Doppler frequency spectrum. These features are called micro-Doppler effect and appear as sidebands around the central Doppler frequency [199].

classification. However, no such highly developed sensors have been available during the course of this work.

Summarizing one can state that the presented EKF-JIPDA is a promising generic fusion approach. It is predestined for fusion approaches that provide information for several applications with different safety requirements. Furthermore, a reliable PoE is valuable for tracking across the sensory blind region and in the blind spot region of the truck to decide if the track from the front shall be trusted and can be used to set up a warning. However, there are some possible improvements. An adequate electronic architecture in the vehicle (e.g., Ethernet-based) and sufficient computational power are basic requirements for the fusion approach, but the main challenge that has to be won is a simple parametrization of the fusion algorithm and the utilized sensors, so that it can be applied in series production.

However, reliable PoEs represent a valuable benefit for situation evaluation modules. New approaches for risk assessment can now be developed based on a novel approach for prediction of the ego vehicle's path. Therefore, the results of this approach will be presented in the following section before the developed risk assessment approach is evaluated.

6.2 Evaluation of the Maneuver Classification and Trajectory Prediction Approach

The development and the evaluation of the novel approach for maneuver classification and trajectory prediction is based on video and CAN recordings. Six different drivers drove the test truck with and without trailer during data recording (about four hours). Defensive and sporty driving styles were tried to be captured. The work of Klanner [200] showed that a single driver exhibits a much higher variance in his maneuver approaches than can be found between the average behavior of different drivers. Thus, the number of different drivers guiding the truck for data acquisition is of little relevance. The data has been labeled manually in Matlab, where a comparison with recorded video data supported the labeling process. The labeling process added points for the beginning and the end of a maneuver. Moreover, temporal reference points have been defined for lane change and turn maneuvers. The reference points have been set to the apexes of the yaw rate signals in case of turn maneuvers and to the inflection points of the yaw rate signals in case of lane changes.

While these two maneuver classes are limited in duration, lane following does not necessarily have a temporal limit and no reference point has been defined for this maneuver class.

About 72 % of all maneuver samples in the recorded data belong to the maneuver class lane following, while about 16 % represent the maneuver class turn. Accordingly, about 12 % of the recorded data samples correspond to the maneuver class lane change. These fractions must not be mistaken for the maneuver distribution on representative routes. As described in Subsection 5.2.2, representative routes have been driven to obtain prior probabilities for the maneuver classes. The fractions for turn maneuvers and lane change maneuvers on these routes have only been 9 % and 4 %, respectively. However, the data analysis and classifier training required a higher number of these two maneuver classes, so that driving of these maneuvers has been forced to increase their fractions. The mean durations and standard deviations of the maneuver classes turn and lane change have been 9.1 s $\pm$ 3 s and 7.4 s $\pm$ 1.6 s. Lane following is not limited in duration and can last several minutes until a new maneuver class comes up. However, the mean duration of this maneuver class was 20 s $\pm$ 35 s in the collected data. Figure 6.13 A illustrates the evolution of

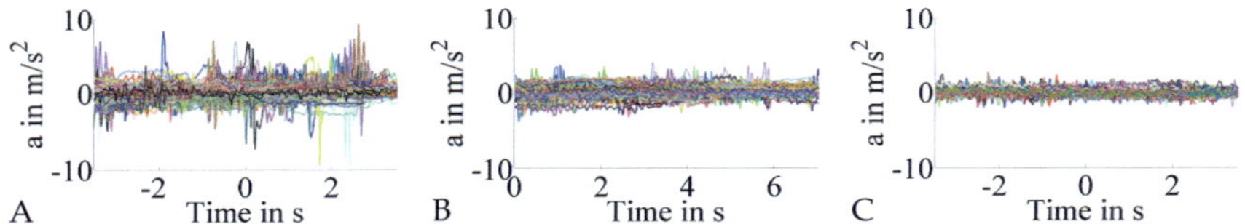

Figure 6.13 Acceleration signals of different maneuver classes without outliers: turn maneuvers (A), lane following maneuvers (B), lane change maneuvers (C).

acceleration signals for the three maneuver classes, where outliers have been excluded for an enhanced visibility. The acceleration signals show the highest variability and the steepest slopes in case of turn maneuvers. The reference points of turn maneuvers are passed at speeds of 15 km/h to 25 km/h, see Figure 6.14 A. Lane change maneuvers are usually driven at higher, constant speeds (see Figure 6.14 C) which is supported by the low acceleration values in Figure 6.13 C. Lane following covers the whole urban speed range (refer to Figure 6.14 B), but the acceleration values take on smaller values than for the maneuver class turn and higher values than lane change maneuvers. The yaw rate signals show the behavior that has

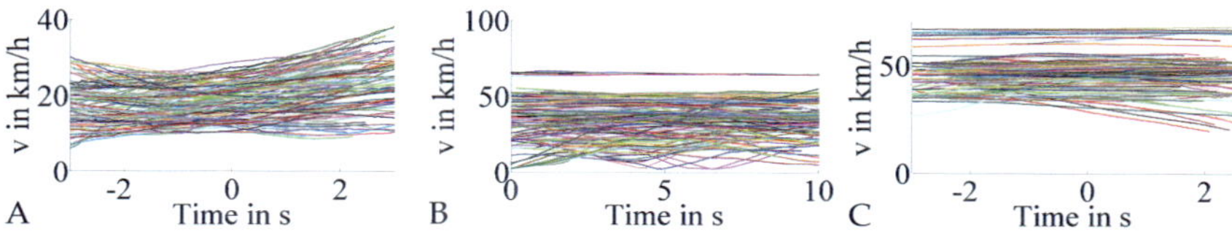

Figure 6.14 Speed signals of different maneuver classes without outliers: turn maneuvers (A), lane following maneuvers (B), lane change maneuvers (C).

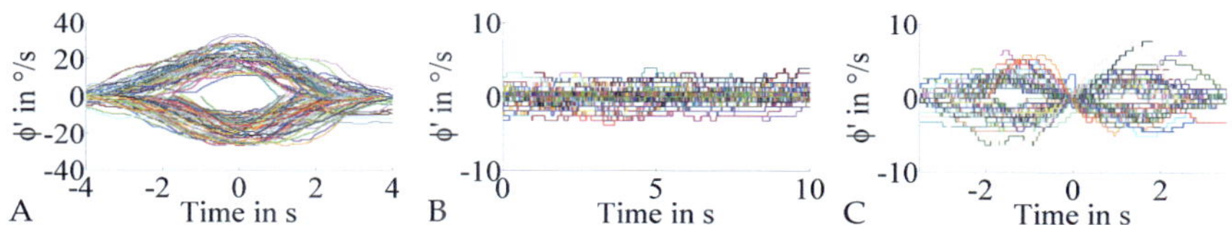

Figure 6.15 Yaw rate signals of different maneuver classes without outliers: turn maneuvers (A), lane following maneuvers (B), lane change maneuvers (D).

already been described in Section 5.2. The yaw rate of a turn maneuver follows a Gaussian curve and reaches the largest values (Figure 6.15 A). Lane following maneuvers show approximately constant and low yaw rate values, see Figure 6.15 B. However, they were within the same value range as the yaw rate signals of lane change maneuvers, although these show a sinusoidal evolution. These signal plots illustrate that the distinction between lane change and lane following is the biggest challenge.

Section 5.2 describes the chosen features for the naive Bayesian classifier. Figure 6.16 shows some of the resulting kernel distributions, where no outliers were removed. The best discrimination is given between turn maneuvers and the other two maneuver classes. Turn maneuvers show the lowest mean speed values but the highest variability in the steering wheel angle and in the acceleration over a duration of 3 s. The speed distributions of the maneuver classes lane change and lane following show local maximums. These are located slightly above the typical speed limits of a truck 30 km/h, 50 km/h and 60 km/h, which appears reasonable.

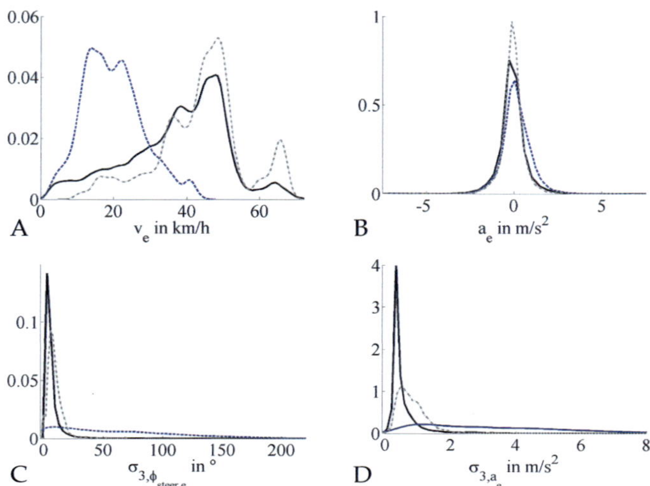

Figure 6.16 Kernel distributions of classifier features obtained from lane following maneuvers (solid black), turn maneuvers (dashed blue) and lane change maneuvers (dash-dotted gray): Speed (A), acceleration (B), standard deviation of the steering angle over the last three seconds (C) and standard deviation of the acceleration over the last three seconds (C).

The features with their corresponding distributions and the computed prior probabilities enable the estimation of the maneuver class. Cross-validation as described in Section 3.3.1 serves the evaluation of the classifier results.

Ten trajectory sequences have been excluded from the data for training in each validation cycle (leave-ten-out). The primarily excluded maneuver sequences are taken as new samples for classification then. A trajectory sequence is represented by a sequence of samples with identical class labels, where one lane following, one lane change or one turning maneuver is included.

For the application in a driver assistance system, it is important to know which maneuver will be driven before it is finished. Therefore, the evaluation relates to the labeled, temporal reference points (turn: yaw rate apex, lane change: yaw rate inflection point). It is checked which

Classified \ Truth	Lane Following	Turn	Lane Change
Lane Following	82%	11%	69%
Turn	9%	89%	3%
Lane Change	9%	0%	28%

Table 6.4 Classifier results one second before the reference point.

fraction of maneuvers the algorithm can classify correctly one second before the apex is reached. Table 6.4 shows the obtained results.

Lane following and turn could be classified correctly most of the time. However, 9 % of the lane following maneuvers have been mistaken for turn maneuvers and 11 % of the turn maneuvers for lane following maneuvers. This can be justified by the labeling procedure. Turn labels have been assigned when the vehicle changed its orientation for more than $45°$. Thereby, the passage between lane following on a curved path and a turn is blurred. However, since the prediction result is similar then, this confusion is acceptable.

The confusion of lane changes and lane following is considerably higher. The speed ranges and the yaw rate ranges are similar, which makes the distinction without availability of lane data extremely challenging. More lane changes have been mistaken for lane following (69 %) than vice versa (only 9 %). This behavior can be explained by the applied prior probabilities that favor lane following maneuvers. It is preferred that a maneuver is classified more likely as lane following than being falsely classified as another maneuver. The standard approach for path prediction assumes a constant acceleration and a constant yaw rate. The new predictor assumes this for lane following as well. Thus, lane following is the neutral maneuver that should be chosen if the feature values do not enable an adequate classification. The new approach will then not perform worse than the standard approach, but it does not provide better results.

Figure 6.17 shows an example for the classification process of a turn maneuver. The blue crosses indicate a correct classification of a turn maneuver in the mid of the sequence. The label turn has been assigned to the samples that are represented by blue circles. The rest of the sequence has been correctly classified as lane following. The dashed gray line in the

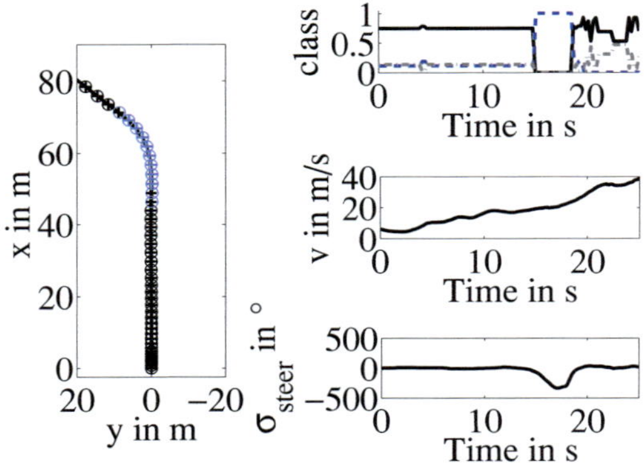

Figure 6.17 Classification result for a turn maneuver and the underlying signals: Circles represent labels and crosses denote the classification result. Blue color is related to turn maneuvers, while black represents lane following.

subgraph at the upper right shows that the probability for a lane change was very low until the end of the sequence.

Figure 6.18 illustrates the classifier's decisions in case of a lane change maneuver. The sequence is classified correctly around the reference point. However, a few samples that have already been labeled as lane following after the end of lane change maneuver are still classified as lane change.

The deviation of the predicted path from the one that is actually driven should be as small as possible in a real application. The standard approach and the new approach are evaluated regarding this aspect. Therefore, each approach projects the path three seconds into the future. This path is compared to the recorded path sequence of this time interval, which is referred to as ground truth. Figure 6.19 and Figure 6.20 visualize the ground truth together with the predicted paths for a turn maneuver and a lane change maneuver, respectively. The maneuvers have been classified correctly. Thereby, the new prediction approach adapts the values of the future yaw rate according to the prediction approach of the maneuver class. The new approach outperforms the standard approach in both cases. The normal distances between the ground truth and the predicted path in certain traveled distances along the path have been computed to obtain a quantitative measure for the prediction accuracy of the newly developed approach and the standard approach.

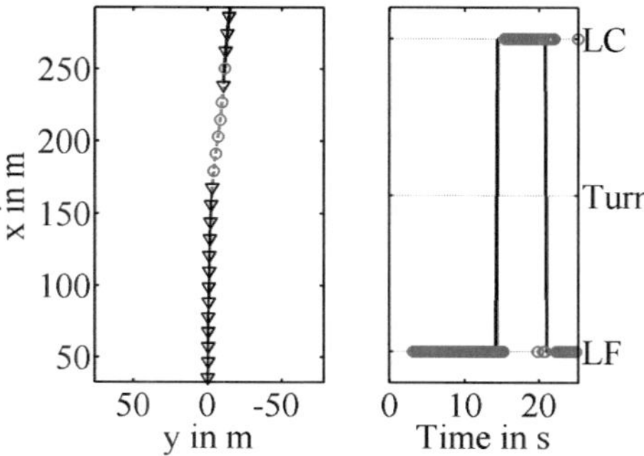

Figure 6.18 Classification result of a lane change maneuver: Gray circles in the left image indicate the classification of a lane change while the dashed gray line represents the section that has been labeled as lane change. The black line in the right image indicates the labeled maneuver class whereas gray circles show the classifier-based decisions.

The box-whisker plots in Figure 6.21 show the 25th and the 75th percentiles of the obtained errors for the new and the standard approach. Figures A (new) and B (standard) represent prediction errors of lane change maneuvers. Visual comparison indicates only little difference between the two approaches. However, a clear difference can be observed for turn maneuvers (see figures C and D).

A Kolmorgorov-Smirnov test has been used to compare the obtained error distributions and to identify if there is a significant difference in the prediction accuracy of the two approaches. Table 6.5 summarizes the obtained p-values for the different maneuver classes. Since the same prediction approach is used in case of a lane following maneuver and most maneuvers are classified correctly as lane following, there is no significant difference for this maneuver class. In case of turn maneuvers, the prediction accuracy of the new approach exceeds the prediction accuracy of the standard approach significantly up to distances of 20 m. The accuracies do not differ significantly at larger distances. Lane change maneuvers could be predicted more accurately using the new approach at distances between 25 m and 30 m, since the new approach takes into account the change in the yaw rate after the reference point. The prediction is the same until the reference point is passed. In total, lane following maneuvers represent the biggest fraction of all maneuvers. Therefore, the

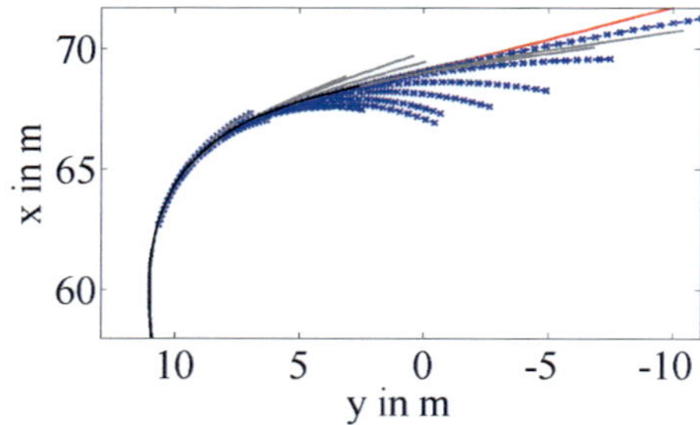

Figure 6.19 Prediction result for a turn maneuver for a prediction horizon of three seconds: The new approach (solid gray line) clearly outperforms the standard approach (dashed blue line) in this situation since it is closer to the actually driven path (red line). The solid black line visualizes the used path history.

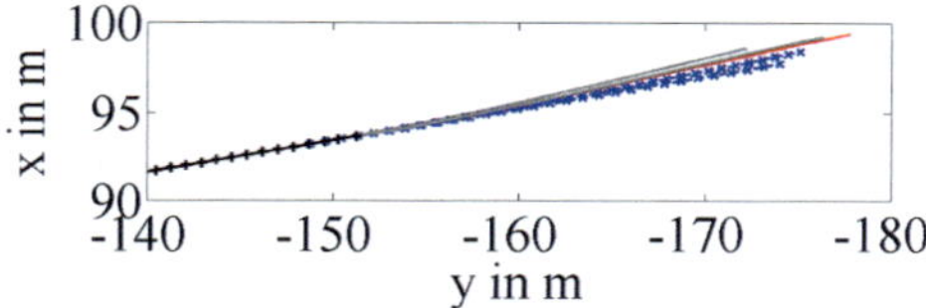

Figure 6.20 Prediction result for a turn maneuver for a prediction horizon of three seconds: The new approach (solid gray line) clearly outperforms the standard approach (dashed blue line) in this situation since it is closer to the actually driven path (red line). The solid black line visualizes the used path history.

prediction accuracy of lane following maneuvers dominates the total result. However, the newly-developed approach outperforms the standard approach in the range up to 15 m.

Moreover, the RMSE values of the path deviations have been determined dependent on the maneuver class and are summarized in Table 6.6. The RMSE values of the two approaches show the biggest difference for turn maneuvers at a distance of 20 m, where the RMSE value is 1.30 m for the standard approach and only 0.79 m for the newly-developed approach. In case of lane change maneuvers, the largest RMSE difference of 0.13 m was computed at a distance of 30 m, where the RMSE value was 0.77 m in case of the standard approach and 0.64 m in case of the new approach.

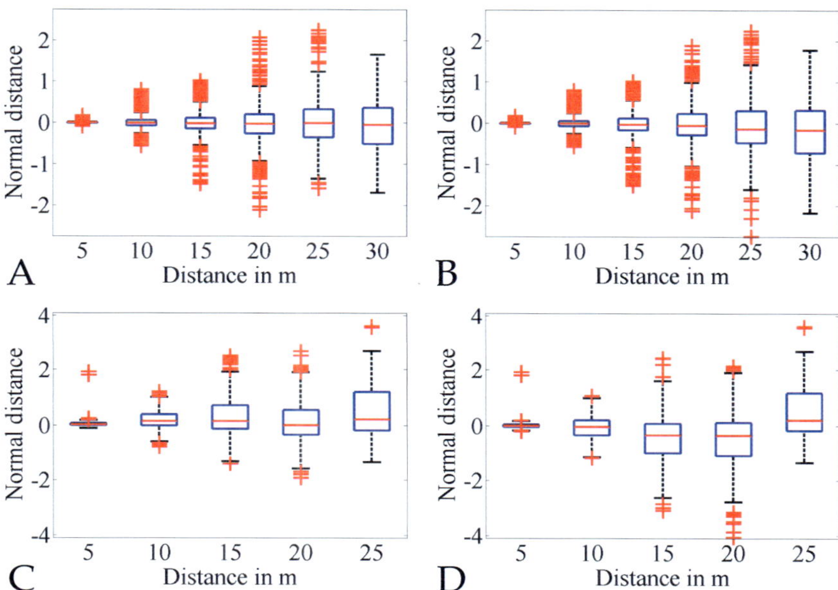

Figure 6.21 Box plots of the normal deviations of the predicted paths from the real paths for lane change maneuvers using the new approach (A) and the standard approach (B) and for turn maneuvers using the new approach (C) and the standard approach (D).

In summary, the new approach outperforms the standard approach. Content of future work could be the improvement of the path prediction. More prior knowledge could be included and the knowledge of the speed range of a turn could be exploited to enable a more accurate prediction. Moreover, other classifiers could be tested for a better distinction between lane change and lane following maneuvers, e.g., SVM or HMM approaches. If additional information from a map and GPS or lane markings were available, this information could be exploited.

Distance	5 m	10 m	15 m
Total	$1.6 \cdot 10^{-5}$	$3.9 \cdot 10^{-7}$	$3.1 \cdot 10^{-5}$
Lane Following	1	1	1
Turn	$7.7 \cdot 10^{-44}$	$1.5 \cdot 10^{-63}$	$3.1 \cdot 10^{-54}$
Lane Change	0.19	0.48	0.40

Distance	20 m	25 m	30 m
Total	0.39	0.86	0.99
Lane Following	1	1	1
Turn	$4.6 \cdot 10^{-10}$	1	–
Lane Change	0.28	$4.9 \cdot 10^{-4}$	0.0198

Table 6.5 P-values of the Kolmogorov-Smirnov test for comparison of the prediction accuracy of the newly developed approach and the standard approach: Values with which the null hypothesis has to be rejected or not. The null hypothesis assumes that the lateral deviations of the predicted paths from the true paths follow the same distribution for the standard approach and the new approach. The evaluation has been performed for different prediction distances along the path.

6.3 Risk Assessment of Vulnerable Road Users

The algorithm for maneuver and path prediction builds an important basis for the presented risk assessment approach that predicts the ego vehicle's behavior along this path. This section discusses the possibilities to evaluate the developed approach for risk assessment and presents the approach's behavior in exemplary situations. In this context, the developed approach is compared to a TTC-based standard approach.

The evaluation of the risk assessment approach is somewhat challenging when real data should be used since there is no ground truth for the quantity *situation risk*. However, one can perform a comparative study between two algorithms, comparing the issuance time of warnings in dangerous situations and the frequency of false alarms in non-dangerous situations.[7] The decision whether or not a road scenario is dangerous could be based on a visual inspection of recorded videos. However, experiments conducted in the course of this work showed that most scenarios were considered much more dangerous when study participants sat in

---

[7]Of course, one does not want to crash naturally moving pedestrians, so that human test persons have to judge situations that never lead to a collision.

Distance		5 m	10 m	15 m
RMSE$_{total}$	new	0.0382	0.1439	0.2567
	standard	0.0386	0.1613	0.3094
RMSE$_{lf}$	new	0.0245	0.0765	0.1778
	standard	0.0245	0.0765	0.1778
RMSE$_{turn}$	new	0.0859	0.3668	0.7647
	standard	0.0878	0.4196	0.9872
RMSE$_{lc}$	new	0.0297	0.1436	0.2791
	standard	0.0307	0.1535	0.3212
Distance		20 m	25 m	30 m
RMSE$_{total}$	new	0.3360	0.4364	0.5328
	standard	0.3820	0.4468	0.5390
RMSE$_{lf}$	new	0.2989	0.4153	0.5222
	standard	0.2989	0.4153	0.5222
RMSE$_{turn}$	new	0.7910	1.2259	–
	standard	1.3007	1.2049	–
RMSE$_{lc}$	new	0.4589	0.5509	0.6482
	standard	0.4684	0.6641	0.7717

Table 6.6 RMSE values of normal distances in meters between the predicted trajectories and the really driven trajectories at a certain distance along the path, where the newly-developed approach and the standard approach have been used for prediction.

the truck than when they watched the video of the situation later. Hence, there seems to be a difference between feeling the acceleration and speed of the vehicle or just seeing the object coming closer in a video. The advantage of video inspection is that several different persons can judge the criticality of the same situation.

Simulated data always contains model assumptions and a deterministic pedestrian behavior, but the advantage of the presented approach is that it is able to model present uncertainty in the road users' behavior. Therefore, simulations might not be able to show the benefit of the presented approach. Simulated data with added noise might already include model assumptions which have an impact on the simulation results. Therefore, real data is used for evaluation in this work and the results of the proposed approach are compared to an established TTC-based standard approach. The following subsection describes this TTC-based approach for risk assessment.

The aim of this work is to show that the assumptions that have been made for the approach are reasonable. The congruence of the system's and human's criticality estimation has to be evaluated in further studies under the guidance of psychologists.

6.3.1 Computation of the Time-to-Collision

The TTC-based algorithm is adapted from the approach in [201] for collision avoidance with vehicles. The approach determines if an object is within the driving corridor[8] (lateral conflict) when the ego vehicle reaches the longitudinal position of the object (longitudinal conflict). The system computes how much time is left to avoid a collision by an intervention, such as braking, acceleration or evasion by steering, if a collision risk is predicted. The corresponding time reserve is referred to as time-to-react. One assumes that the ego vehicle drives along a path with constant curvature and keeps its acceleration a_e constant. No uncertainty is considered for the states and the behavior of the objects. The curved path of the ego vehicle builds the x-axis of the coordinate system in which objects and the ego vehicle are represented by rectangles. Thus, all detected object positions have to be transformed to the path-aligned coordinate system first. Hillenbrand [201] assumes a constant acceleration for objects since only cars are considered. Here, a constant speed is assumed when the detected object is a pedestrian. Finally, one can formulate time reserves for the computation of a binary collision risk. These time reserves and required parameters are visualized in Figure 6.22. The time-to-appear (TTA) is the time when an object appears in the driving corridor of the ego vehicle:

$$
\mathrm{TTA} = \begin{cases}
-\dfrac{y_\mathrm{obj}-0.5\cdot w_\mathrm{obj}-0.5\cdot w_\mathrm{e}}{v_\mathrm{y,obj}}, & \text{if } \left(y_\mathrm{obj} - \dfrac{w_\mathrm{obj}}{2} > \dfrac{w_\mathrm{e}}{2}\right) \wedge \left(v_\mathrm{y,obj} \neq 0\right), \\
\dfrac{y_\mathrm{obj}+0.5\cdot w_\mathrm{obj}+0.5\cdot w_\mathrm{e}}{v_\mathrm{y,obj}}, & \text{if } \left(y_\mathrm{obj} + \dfrac{w_\mathrm{obj}}{2} < -\dfrac{w_\mathrm{e}}{2}\right) \wedge \left(v_\mathrm{y,obj} \neq 0\right), \\
0, & \text{if } |y_\mathrm{obj}| + \dfrac{w_\mathrm{obj}}{2} < \dfrac{w_\mathrm{e}}{2}, \\
\infty, & \text{else,}
\end{cases}
\tag{6.12}
$$

---

[8] The driving corridor is defined as tube of defined width (e.g., half vehicle width) around the predicted path of the ego vehicle.

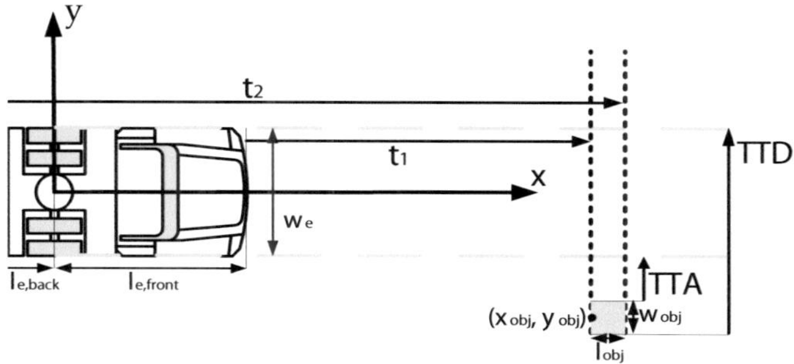

Figure 6.22   Illustration of different time reserves for the TTC computation.

where w_{obj} represents the width of the object. The width w_e of the ego vehicle defines the width of the driving corridor. The positions x_{obj} and y_{obj} correspond to the object position in the curve-aligned coordinate system. The time-to-disappear (TTD) is the time until an object will have left the driving corridor:

$$
\text{TTD} = \begin{cases}
\text{TTA} + \dfrac{w_e + w_{\text{obj}}}{|v_{y,\text{obj}}|}, & \text{if } (0 < \text{TTA} < \infty) \wedge (v_{y,\text{obj}} \neq 0), \\
\dfrac{0.5 \cdot w_e + 0.5 \cdot w_{\text{obj}} - y_{\text{obj}}}{|v_{y,\text{obj}}|}, & \text{if } (\text{TTA} = 0) \wedge (v_{y,\text{obj}} \neq 0), \\
\infty, & \text{if } (\text{TTA} = 0) \wedge (v_{y,\text{obj}} = 0), \\
0, & \text{else.}
\end{cases}
\tag{6.13}
$$

The time-to-collision (TTC) checks when the longitudinal positions of the object and the ego vehicle intersect for the first time. Therefor, the two auxiliary times t_1 and t_2 are computed

$$
t_1 = \begin{cases}
\dfrac{-\Delta v + \sqrt{\Delta v^2 - 2 a_e (l_{e,\text{front}} - x_{\text{obj}})}}{a_e}, & \text{if } (\Delta v > 0) \wedge (a_e \neq 0), \\
\dfrac{x_{\text{obj}} - l_{e,\text{front}}}{\Delta v}, & \text{if } a_e = 0, \\
0, & \text{if } (x_{\text{obj}} + l_{\text{obj}} + l_{e,\text{back}} \geq 0) \\
& \quad \wedge (x_{\text{obj}} - l_{e,\text{front}} \leq 0),
\end{cases}
\tag{6.14}
$$

where the assumption $\Delta v = (v_e - v_{x,\text{obj}}) > 0$ is reasonable since the ego vehicle is driving in forward direction and the vehicle speed exceeds the speed of a pedestrian in the speed range of the system. The parameter l_{obj}

represents the object length, while $l_{e,\text{front}}$ is the distance from the origin of the coordinate system to the front of the ego vehicle.

$$t_2 = \begin{cases} \dfrac{-\Delta v + \sqrt{\Delta v^2 + 2a_e\left(x_{\text{obj}} + l_{\text{obj}} + l_{e,\text{back}}\right)}}{a_e}, & \text{if } (\Delta v > 0) \wedge (a_e \neq 0), \\ \dfrac{x_{\text{obj}} + l_{\text{obj}} + l_{e,\text{back}}}{\Delta v}, & \text{if } a_e = 0, \end{cases} \tag{6.15}$$

where $l_{e,\text{back}}$ is the distance from the origin to the rear-end of the ego vehicle. The TTC is then given by

$$\text{TTC} = \begin{cases} \infty, & \text{if } (\text{TTD} \leq t_1) \vee (\text{TTA} > t_2), \\ \max(t_1, \text{TTA}), & \text{else.} \end{cases} \tag{6.16}$$

If $\text{TTC} \rightarrow \infty$, no collision is expected. If the TTC is less than 3 s, the system would consider to issue a warning to the driver.

The TTC approach is compared to the newly developed approach. Therefore, some exemplary situations and the approaches' outcomes are presented in the following subsection.

6.3.2 Collision Risk between the Ego Vehicle and Pedestrians in Exemplary Situations

Section 5.3 described observations for pedestrian behavior. Values for maximum acceleration and speed have been measured and have been compared to the literature. The basic assumptions for pedestrian behavior and motion parameters have been retrieved from these studies. For the discretization of the state space and the input space, one has to find a compromise between model accuracy and computational complexity. Although theoretical pedestrian speeds higher than 6 m/s can be reached, these speeds are very rare in traffic. If these speeds really occur, it is unlikely that the corresponding pedestrians are tracked. Therefore, the speed range of pedestrians is limited to 6 m/s for the simulations.

The state space has been discretized to 124 valid position intervals for motorized road users, 97 for bicyclists and to 88 valid position intervals for pedestrians, as can be derived from Table 5.1. The positions and the speed along the path build the two-dimensional state space. The speed range is split into 20 intervals for vulnerable road users and into 22 intervals for motorized vehicles. Six intervals model the input space for Markov chain abstraction that represents acceleration and deceleration. The Markov chain has been abstracted for six time intervals in the range between $[0\,\text{s}, 3\,\text{s}]$, where each interval is divided into ten intermediate time steps for abstraction.

The performance of the approach is visualized for situations that are related to the accident situations in Subsection 2.3.1. Figure 6.23 illustrates the reachable sets of the ego vehicle and of a pedestrian for prediction intervals of 0.5 seconds up to a prediction horizon of three seconds. The

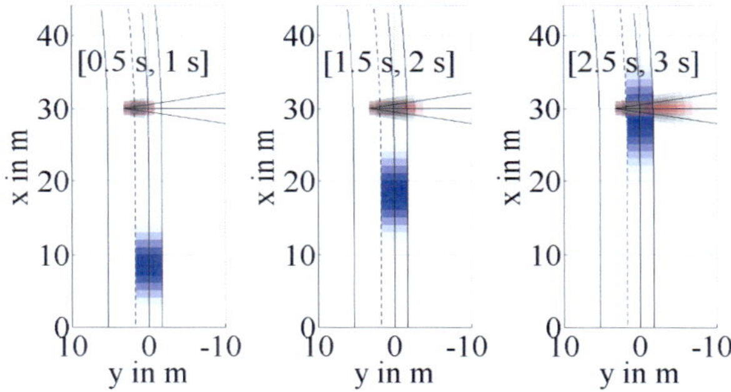

Figure 6.23 Illustration of reachable sets of the ego vehicle and a pedestrian for three prediction intervals.

pedestrian is detected at the left side of the lane in 30 m longitudinal distance with a lateral offset of 4 m. The measured absolute speed mean of the pedestrian is 1.3 m/s. The PoE of the pedestrian is estimated to be 0.97. The pedestrian is expected to cross the lane in normal direction, which has been retrieved from the motion orientation of the last measurement cycles. This path obtains a probability of 0.6. The state variance of the lateral position from the environment perception provides the initial position uncertainty along the path. The speed uncertainty is computed analogously. Two additional paths are assumed for the pedestrian to account for the possibility that the estimated direction was not correct. These two paths differ from the first path in orientation by about $9°$ and get assigned probabilities of 0.2 each. The initial state intervals along the paths represent the uncertainty of the initial state. The first four elements of the principal diagonal of the state covariance matrix of the detected pedestrian are used to compute a 3σ-ellipse of the position and a 3σ-ellipse of the speed.[9] The values for x, y, v_x and v_y in the example are summarized in the principal diagonal $\mathrm{diag}(\Sigma_{\mathrm{obj}}) = (0.4\,0.15\,0.3\,0.15)$.

---

[9] The computation of the ellipsoid points is based on
$x_{\mathrm{Ellipse}} = \sigma_x \cos(\arctan(\sigma_x \tan(\phi)/\sigma_y))$ and $y_{\mathrm{Ellipse}} = \sigma_y \sin(\arctan(\sigma_x \tan(\phi)/\sigma_y))$.

Thus, the interval sets representing the initial states can be computed. The deviation from the path is obtained from the position rectangles as well where the deviation is oriented orthogonal to the path. The ego vehicle is expected to certainly follow the lane (100%). Its current speed is 36 km/h, where the maximum inaccuracy must not exceed 1 km/h. The algorithm expects that the ego vehicle moves on at an approximately constant speed in this situation, so that its initial input distribution is set to $(0\ 0\ 0.5\ 0.5\ 0\ 0)^{\mathrm{T}}$.

The lateral offset of 4 m is quite large, so that it is expected that the pedestrian is still outside the predicted driving corridor of the ego vehicle. The probability that the pedestrian will decelerate and stop at the street's edge to watch out is relatively high. It is estimated as high as the probability that the pedestrian will continue walking at the same speed. Therefore, the initial probability vector for the initial distribution of inputs is set to $(0.5\ 0\ 0.25\ 0.25\ 0\ 0)^{\mathrm{T}}$. When the pedestrian approaches the ego lane and the lateral offset decreases, e.g., to 3 m, the probability that the pedestrian will keep on walking increases compared to the probability that he will stop, as he has probably stepped onto the street. Based on the initial setting, the algorithm determines the probability of a collision with the pedestrian for the time interval $[2.5\text{ s}, 3\text{ s}]$ to 8 %. The standard approach would compute a TTC of 2.97 s if the mean values of the estimated object state are used. However, the measurement data may not be accurate and the real lateral position of the pedestrian could be 4.45 m and his real speed could be 0.85 m/s based on the spatial uncertainty. No collision would be expected for these values. If the detected object has just been initialized, so that the estimated speed is still very uncertain and the mean speed value is still rising, the speed value that is provided as estimated mean could be 0.85 m/s, although the real pedestrian speed is 1.30 m/s. Furthermore, the mean lateral position could be over-estimated to 4.45 m. Thus, the TTC-based system would not expect a collision even though the situation might be critical. In contrast, the developed approach based on reachable sets still computes a collision probability of 7 % for the time interval $[2.5\text{ s}, 3\text{ s}]$ with these mean values and the covariance matrix Σ_{obj} from above. The comparison has been performed for exemplary values, but situations with other values demonstrate as well that the TTC approach over-estimates or under-estimates the collision risk. It cannot accurately determine the time point of a collision because it does not take into account the known uncertainties.

The second situation in Figure 6.24 shows a T-crossing. A pedestrian with an PoE of 99 % follows the lane, whereas the ego vehicle is expected to turn into the street on the right with a probability of 90 % and to follow the lane only with 10 %. The pedestrian is detected within the longitudinal position interval $[15 \text{ m}, 16.2 \text{ m}]$ with a speed range of $[0.7 \text{ m/s}, 1.6 \text{ m/s}]$. The model assumes that he will keep his speed. The ego vehicle is initially running at 5.6 m/s and it is expected to decelerate in case of the turn, so that the initial input distribution is set to $(0.1 \ 0.5 \ 0.35 \ 0.05 \ 0 \ 0)^{\text{T}}$. If the vehicle follows the lane, the input probability is equally distributed. The maximum collision probability is predicted to be 7 % in the time interval $[1.5 \text{ s}, 2 \text{ s}]$. The TTC-based approach

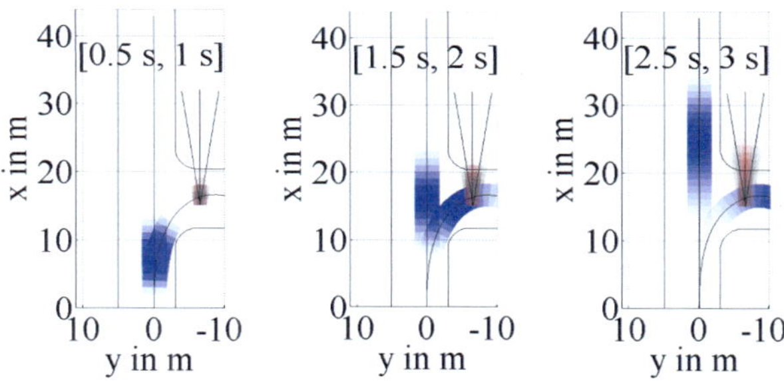

Figure 6.24 Illustration of reachable sets of the ego vehicle and a pedestrian during a turning maneuver of the ego vehicle. The maximum collision probability is predicted to be 7 % in the time interval $[1.5 \text{ s}, 2 \text{ s}]$.

does not expect a collision in this situation. The turning maneuver cannot be modeled by a path with constant curvature. If the constant curvature assumption is applied with the initial curvature, one obtains the lane following path for which no collision risk is predicted by both approaches. Therefore, the standard approach should be extended in such way that it uses several path predictions to handle these situation. However, the extension of a TTC-based approach is not content of this work.

Finally, both approaches shall be compared using a measurement sequence of a situation where two pedestrians cross the lane. The ego vehicle approaches the pedestrians from a left curve, so that the pedestrians are detected for the first time at a short distance. Consequently, the state uncertainty is still relatively high. Moreover, the sequence has been recorded when the sun was in low altitude, so that the camera's detection performance decreases and available HOG detections have relatively low confidence values. The ego vehicle approaches the pedestrians at an initial speed of 28 km/h with a mean acceleration of -0.25 m/s^2. No collision took place and the pedestrians did not have to start running to avoid a collision. Figure 6.25 shows the evolution of the computed crash probabilities where a lane following maneuver is predicted for the ego vehicle. The approach based on reachable sets predicts a maximum crash risk of 20 %. The vertical lines represent the points of time for which the TTC-based approach predicts a collision.

The computation of the collision risk with one pedestrian took 30 ms on average but never longer than 100 ms.[10] Thus, under the assumption that an update rate of 100 ms is sufficient, the approach would be applicable in real-time if the approach was implemented in C++, the number of considered pedestrians was limited, or the computation of the conditional crash probabilities with single pedestrians was parallelized.

The TTC-based crash prediction would sporadically set alarms at different time points and there is no collision in the end, while the newly-developed approach predicts the maximum collision risk around one time point and a maximum of 20 %. If the TTC-based system includes a heuristic that checks the stability of the prediction, valuable time for reaction would be lost in case of a serious collision risk. The shape of the crash probabilities evolution approximates the density of a normal distribution around 9.5 s. Which threshold values of the crash probability suffice to initiate an alarm or an active intervention into the vehicle dynamics would have to be analyzed in extensive studies with several drivers in real traffic.

Summarizing, the newly-developed approach is able to handle uncertainty in the state and in the existence of a tracked object. Moreover, it is able to handle uncertainty in the prediction of the ego motion and includes different maneuver classes that cannot be modeled by path predictions based on the assumption of a constant curvature. However, the computational complexity of the standard approach is significantly lower than for the new approach that is based on the intersection of reachable sets for time intervals. On the other hand, the new approach provides more stable temporal predictions for a collision than the TTC-based standard approach and the application of time intervals avoids that time points of a potential collision are missed (tunnel effect). However, if the TTC was computed for different configurations of object position and object speed based on the estimated uncertainty, one would obtain more stable values as well. The availability of a non-binary collision risk could

---

[10]The computations were performed with a 32-bit version of Matlab on an Intel processor (Intel(R) Core (TM) 2Duo CPU T9400 @2.53GHz).

enable graded system interventions based on the collision risk. For example, the system could cut off the engine and pre-condition the brakes in case of low crash probabilities to gain time without disturbing the driver, while high crash probabilities could induce an emergency braking after a warning phase. Finally, a discrete value has to be computed to enable the start of the system reaction.

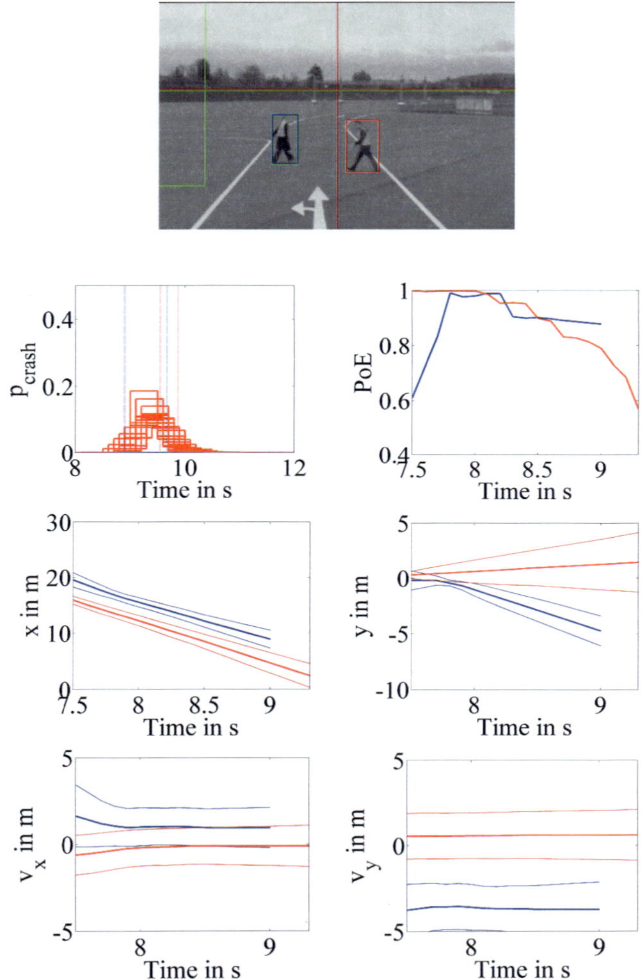

Figure 6.25 Evolution of the predicted crash probabilities, the pedestrians' positions and speeds as well as of the corresponding PoEs in a situation with two pedestrians crossing the lane: The initial speed of the ego vehicle is 28 km/h. The sensory uncertainty is relatively high, so that the risk of a crash is estimated to be 20 % at maximum.

7 Conclusion and Outlook

7.1 Conclusion

New solutions to problems regarding environment perception and situation evaluation in ADAS of commercial vehicles have been presented.

An EKF-JIPDA filter has been developed, implemented in a C++-based framework and was parametrized to track pedestrians from a truck in real-time using a monocular camera and two radars at the front and one radar at the right side of the truck. The tracking results of the newly-developed approach have been compared to the results of an EKF-GNN filter. Two parameter sets of the EKF-GNN have been used for the evaluation with measurement data of various urban scenarios showing several pedestrians. The EKF-JIPDA outperforms the EKF-GNN significantly regarding the detection performance in crowded scenes, whereas only little difference was observed in the state estimation of both filter approaches.

The EKF-JIPDA and the EKF-GNN have been used to track pedestrians across a sensory blind region and in distinct FOVs where the state estimation of the EKF-JIPDA is more accurate. Tracking across the sensory blind region makes camera-specific information — such as the height or the type classification of an object — from the vehicle's front available in the sensory blind region and in the second FOV, where only a radar tracks the objects. Multiple objects can be tracked reliably over durations of several seconds with one ID per physical object and the integrated existence estimation provides valuable information for subsequent situation evaluation modules of the ADAS.

Pedestrian detections have been filtered from point sets provided by a laser scanner using linear regression lines and a DBSCAN algorithm. These detections have been used as reference detections to measure sensor-characteristic properties of the radar sensors and the camera such as the spatial accuracy of the measurements, the spatial detection rate and measurement-dependent inference probabilities that provide evidence if a measurement results from a relevant object class (pedestrian) or not. The results have been used to parametrize the filter approaches. Moreover, the filtered detections from the laser scanner served as reference for the evaluation of the spatial filter accuracy. Manually labeled images enabled the evaluation of the detection performance of the developed and implemented filter approaches.

For situation evaluation, a new approach for the early classification of a driving maneuver of the ego vehicle using the LCS method and a Bayesian classifier has been developed. Furthermore, a possibility has been shown to utilize the classification result for long-term trajectory prediction based on the current speed and the typical yaw rate evolution of the corresponding maneuver class. The performance of the developed classifier has been trained and cross-validated using manually labeled trajectories obtained from recorded CAN signals. Several drivers drove a truck with and without trailer through the urban environment in the south-west of Germany for several hours to record the signals. The prior probabilities of the classifier have been set in such way that a maneuver is classified as lane following maneuver in case of uncertainty, since this is the maneuver with the lowest risk compared to a standard approach based on the assumption of constant speed and constant yaw rate. The developed approach for maneuver prediction was compared to the named standard approach. The prediction accuracy of the new approach outperforms the standard approach.

An important contribution of this work is the development of a new approach to assess the collision risk of the ego vehicle with other road users, especially pedestrians. Therefor, the motion behavior of pedestrians in traffic has been analyzed using information from the literature and by performing own exploratory studies. The paths of the ego vehicle and of pedestrians have been predicted in different situations. Stochastic reachable sets of the road users along their paths have been computed for time points and time intervals. Monte-Carlo simulation and subsequent Markov chain abstraction enable an efficient online computation of the stochastic reachable sets of the different road users. The intersection of the stochastic reachable set of the ego vehicle with the stochastic reachable set of a detected road user provides the crash probability. This crash probability is then taken as the newly-developed risk measure and is compared to a TTC-based standard approach for risk assessment. In contrast to the existing approach, the new approach can handle sensory uncertainty as well as uncertainty in the future motion of the road users and performs more stably in such scenarios.

7.2 Outlook

The generic sensor fusion framework for pedestrian tracking still provides a number of false alarms in dense environments, since the radar can only barely distinguish between relevant objects like pedestrians and other weakly reflecting objects. The developed approach could be extended by the ideas of Munz [65]. He utilizes Dempster's and Shafer's theory of evidence to discern between non-objects, objects and relevant object classes. Thereby, sensors that cannot provide a sensory inference probability for the object class can be modeled as such. Moreover, future radars with higher resolutions and additional computational power will classify pedestrians based on the motion of their extremities, so-called micro-Doppler classification. Thus, future radars might provide sensory inference probabilities.

Discrimination between different objects with little distance is not always possible due to a limited sensor resolution. Only one detection is obtained for two objects then. This fact could be explicitly modeled by considering the merging and the separation probability of measurements from low-resolution sensors as in [96, 202] for radars. This is especially interesting for scenarios where several pedestrians cross the street from different sides and temporarily occlude each other or pedestrians have been located close to objects like parked cars at the road edge.

A constant-speed model has been chosen to model the motion of pedestrians. An pessimistic assumption for the model uncertainty had to be made to model highly dynamic scenarios. The motion models that fit best for the present situation could be used if multi-instance filters were integrated. Furthermore, future work should investigate the filter performance if an unscented Kalman filter (UKF) was implemented instead of an EKF for state estimation. Especially, in case of strong non-linearities the UKF enhances the state estimation compared to the EKF without requiring the determination of a Jacobian matrix.

The dynamic pitch angle of the driver's cabin impacts the accuracy of the camera's state estimation. Computation of power flow could enable the estimation of the vehicle's pitch angle and could thereby improve the state estimation. Additional heuristics could filter the obtained pedestrian tracks to obtain more reliable pedestrian detections. Moreover, the tracking of the EKF-GNN could be improved by integrating some components of the EKF-JIPDA into the existence estimation and the track management, such as the likelihood probability or the availability of HOG detections with a high sensory inference probability. One has to find a simple and robust way to set the sensor parameters in different trucks to enable an application of the EKF-JIPDA approach in series production. Additional evaluations of the estimators' state accuracy should be performed with a differential GPS platform for pedestrians. The EKF-JIPDA approach could also be valuable to track objects across sensory blind regions in surveillance or for object tracking from multiple moving sensor platforms as in car-to-car communication, where the FOVs do not necessarily overlap.

Several thousand labeled measurement samples have been used to evaluate the EKF-JIPDA in comparison to the EKF-GNN in this work. However, this is a very small number in comparison to data from 1000000 km that have to be driven with a low number of false alarms before a tracking approach will be finally integrated into a truck for series production. Therefore, the database with labeled samples should be enlarged for further developments. Each system level has to limit its false alarms to a minimum number. The number of false detections from the sensors, the number of false positives of the fusion approach and the number of false alarms from an application have to be kept at a sufficiently low level.

The presented maneuver classification approach performed well for the distinction between lane following maneuvers and turn maneuvers. However, additional approaches should be evaluated, especially, to improve the classification of lane change maneuvers. For instance, HMM or SVM algorithms could be used. The prediction performance strongly depends on the estimation of the correct section in the maneuver sequence. Further developments could include Gaussian mixture models [148] for each maneuver class to improve the prediction performance and to make the prediction more independent of speed and road geometry.

The approach for risk assessment has been developed based on assumptions obtained from studies on the motion behavior of pedestrians and the ego vehicle. So far, no pedestrian interaction has been considered, but future developments could adapt pedestrian behavior based on the environment. Although measurement data was used to compare the novel approach to an established approach, a study in real traffic with real pedestrians and several drivers should be performed to obtain an ROC curve for the false alarm rate based on the crash probabilities and to compare the computed crash probabilities to human risk estimates. The distributions of the crash probabilities could be used to build a hazard map to guide the driver's attention to 'hot spots'. Moreover, the results could be applied to the development of emergency braking systems.

A Appendix

A.1 Selected Proprioceptive Sensors

A MEMS **accelerometer** consists of movable proof mass with plates at the outside creating two combs connected at the closed side. A mechanical suspension system with known stiffness attaches this movable plate to a reference frame. The springs should have a significantly lower stiffness in the measurement axis compared to the other axes. There are two fixed plates around each inner plate where one is connected to the negative supply voltage and one to the positive. Thereby, one obtains two capacitor constructions around each inner plate. Acceleration induces a deflection of the proof mass (inner plates) which can be measured as capacitive difference.

Gyroscopes measure the orientation based on the principles of angular momentum. Many yaw rate sensors are based on the following principle. An electromagnetic comb drive induces oscillation of a pair of masses with equal amplitude but in opposite directions. Vehicle rotation around one of the sensors' in-plane axes (e.g., the vertical axis) leads to a Coriolis force that creates an orthogonal vibration. The moving masses lift and induce a capacitive change that can be detected with capacitive electrodes under the mass. The measurable Coriolis force is proportional to the turn rate. More details about gyroscopes and accelerometers can be found in [203].

Wheel speed encoders are used to determine the rotational wheel speed and enable the computation of the vehicle speed in combination with the wheel circumference. A rotating wheel creates a magnetic field at the sensor, so that it is based on the magneto-resistive principle or Hall principle. Either one uses a magnetic multi-pole wheel to create a periodically varying magnetic field or a gear wheel made of ferromagnetic material and an external magnet.

The **steering wheel angle** is measured around the steering column. The principle is based on two gear wheels that differ in the gear transmission ratio to the hub of the steering column by two. Magnets at the gear wheels induce a resistance change in the GMR (Giant Magneto Resistive) elements located at the opposite side that is proportional to the angle. The phase-shifted voltage curves and the Vernier principle enable an unambiguous determination of the initial position [204].

A.2 Parameters of Pedestrian Motion

		Speed Range in m/s	Mean Speed in m/s	a_{start} in m/s^2	t_{start} in s
Walking	[48]	1.19 – 1.68	1.42	1.34 – 4.86	1.72
	[46]	1.25 – 1.5	1.38		
-Young Pers.	[47]				2 – 3
	[45]		1.48		
-Old Pers.	[47]		0.9		2.5 – 3.75
	[45]		1.32		
-Groups	[46]	1 – 1.5		0.25 – 0.5	
Fast Walking	[48]	1.62 – 2.34	1.98		
Jogging	[48]	2.31 – 3.96			
-Men	[48]		3.20		
-Women	[48]		3.03		
Running	[48]	3.40 – 6.36			
-Men	[48]		4.99		
-Women	[48]		4.51		
Red Traffic Light	[47]		1.6		
Zebra Crosswalk	[46]	1.75 – 2.5	2		

Table A.1 Results from studies regarding pedestrian speed analysis. Study participants have been instructed to perform a specific walking behavior in [48], while the motion of naturally moving pedestrians has been recorded in [45–47]. a_{start} represents a pedestrian's acceleration from standstill to constant end-speed. t_{start} refers to the corresponding time that the pedestrian needs to reach his constant end-speed. It is measured from the time point when the traffic light signal turns green in [46, 47] and thus, it includes the reaction time of the pedestrians. In [48], the time does not include the reaction time. Older persons are considered as persons that appeared to be older than 65 years.

A.3 Cross-Validation

Hold-out cross-validation (simple cross-validation) and N_s-fold cross-validation are popular methods to select the model with the best performance from a finite set of models $\mathcal{M} = \{M_1, ..., M_{N_p}\}$. Moreover, the algorithms can help to evaluate of a single model or algorithm. The principle is taken from [69].

In the first case, a training set $\mathcal{S}$ is randomly split into the subset $\mathcal{S}_{\text{train}}$ (usually 70 % of $\mathcal{S}$) and the hold-out cross-validation subset $\mathcal{S}_{\text{cv}}$. Then, each model M_l is trained on $\mathcal{S}_{\text{train}}$ to get a hypothesis $h_{\Theta,l}$. Finally, the hypothesis function $h_{\Theta,l}$ with the smallest number of misclassifications on the examples of the cross-validation set $\mathcal{S}_{\text{cv}}$ is selected. Testing hypothesis function $h_{\Theta,l}$ on the cross-validation set provides a better estimate of the true generalization error. The disadvantage of this method is the waste of 30 % of the labeled data.

The N_s-fold cross-validation method holds out less data each time, but is computationally more expensive. The training set $\mathcal{S}$ of size N_t is randomly split into N_s disjoint subsets $\mathcal{S}_1, \ldots, \mathcal{S}_{N_s}$ with N_t/N_s training samples each. Then, each model M_l is trained on the subsets $\mathcal{S}\backslash\mathcal{S}_m$ with $m = 1, \ldots, N_s$ to get some hypothesis function $h_{\Theta,lm}$. Next, the test of the hypothesis function $h_{\Theta,lm}$ on the remaining subset $\mathcal{S}_m$ provides the corresponding cross-validation error $\hat{\epsilon}_{\mathcal{S}_m}(h_{\Theta,lm})$. The mean of these errors $\frac{1}{N_s}\sum_{m=1}^{N_s}\hat{\epsilon}_{\mathcal{S}_m}(h_{\Theta,lm})$ reveals the estimated generalization error of model M_l. Finally, one picks the model M_l with the lowest estimated generalization error and retrains the model on the entire training set $\mathcal{S}$ leading to the final hypothesis output h_Θ. A special type of this method is called leave-one-out cross-validation where the number of training subsets N_s corresponds to the number of training examples N_t.

A.4 Interval Computations

Operations on two intervals $\underline{\bar{a}} = [\underline{a}, \bar{a}] \in \mathcal{I}$ and $\underline{\bar{b}} = [\underline{b}, \bar{b}] \in \mathcal{I}$ are denoted by

$$\underline{\bar{a}} \circ \underline{\bar{b}} = \{a \circ b \,|\, a \in \mathbf{a}, b \in \mathbf{b}\}, \tag{A.1}$$

where $\circ$ represents the operator. Addition and multiplication of intervals are operations that should be briefly presented, since they are utilized in this work to represent state sets of road users:

$$\overline{\underline{a}} + \overline{\underline{b}} = [\underline{a} + \underline{b}, \overline{a} + \overline{b}], \tag{A.2}$$

$$\overline{\underline{a}} \cdot \overline{\underline{b}} = [\min(\underline{ab}, \underline{a}\overline{b}, \overline{a}\underline{b}, \overline{a}\overline{b}), \max(\underline{ab}, \underline{a}\overline{b}, \overline{a}\underline{b}, \overline{a}\overline{b})]. \tag{A.3}$$

If one applies the enumerated formulas for the range computation of a function, one has to ascertain that each variable occurs only once in interval computations to obtain the exact solution. Otherwise, the exact solution is included, but cannot be guaranteed. For example, the function $\overline{\underline{c}} = \overline{\underline{a}} \cdot \overline{\underline{b}} + \overline{\underline{a}}$ can be reformulated according to $\overline{\underline{c}} = \overline{\underline{a}} \cdot (\overline{\underline{b}} + 1)$. Let be $\overline{\underline{a}} = [-5, -2]$ and $\overline{\underline{b}} = [-1, 5]$, then one gets $\overline{\underline{c}} = [-30, 3]$ in the first case and $\overline{\underline{c}} = [-30, 0]$ in the latter. This results from the fact that the values of the operands may take any value within the specified interval regardless of previous occurrences for each evaluation of an interval operation. Thus, different values of the same operand lead to the minimum and maximum values of the corresponding interval operations, although the same operand is not allowed to have different values at the same time. Expressions with single use of variables are referred to as single-use expressions (SUE) [205]. The problem of over-approximative results for non-single-use expressions is also referred to as the dependency problem in the literature [206].

Interval arithmetic will also be applied to interval matrices $\overline{\underline{\mathbf{A}}} = [\underline{\mathbf{A}}, \overline{\mathbf{A}}] \in \mathcal{I}^n$ where $\underline{\mathbf{A}} \in \mathbb{R}$ and $\overline{\mathbf{A}} \in \mathbb{R}$ are the left and right limit of the interval matrix, so that $\overline{\underline{\mathbf{A}}} = [\underline{\mathbf{A}}, \overline{\mathbf{A}}]$. They are required to consider uncertain system matrices $\overline{\underline{\mathbf{A}}}$. The power of an interval matrix cannot be formulated as a SUE in general. The simplest case of computing an interval matrix product $\overline{\underline{\mathbf{C}}} = \overline{\underline{\mathbf{A}}} \cdot \overline{\underline{\mathbf{B}}}$ where $\overline{\underline{\mathbf{A}}} \in \mathcal{I}^{m \times n}$ and $\overline{\underline{\mathbf{B}}} \in \mathcal{I}^{n \times m}$, is done based on single matrix elements resulting in the SUE: $\overline{\underline{\mathbf{C}}}_{ij} = \sum_{k=1}^{n} \overline{\underline{\mathbf{A}}}_{ik} \overline{\underline{\mathbf{B}}}_{kj}$. Thus, one obtains the exact results when applying interval arithmetic.

Correspondingly, the square of a matrix $\overline{\underline{\mathbf{C}}} = \overline{\underline{\mathbf{A}}}^2$ can be written as SUE as well:

$$\overline{\underline{\mathbf{C}}}_{ij} = \begin{cases} \overline{\underline{\mathbf{A}}}_{ij}(\overline{\underline{\mathbf{A}}}_{ii} + \overline{\underline{\mathbf{A}}}_{jj}) + \sum_{k:k\neq i, k\neq j} \overline{\underline{\mathbf{A}}}_{ik} \overline{\underline{\mathbf{A}}}_{kj}, & \text{if } i \neq j, \\ \overline{\underline{\mathbf{A}}}_{ij}^2 + \sum_{k:k\neq i, k\neq j} \overline{\underline{\mathbf{A}}}_{ik} \overline{\underline{\mathbf{A}}}_{kj}, & \text{if } i = j. \end{cases} \tag{A.4}$$

There is no SUE for the multiplication of three interval matrices. Furthermore, matrix multiplication is not associative when using interval arithmetic: $(\overline{\underline{\mathbf{A}}}\overline{\underline{\mathbf{B}}})\overline{\underline{\mathbf{C}}} \neq \overline{\underline{\mathbf{A}}}(\overline{\underline{\mathbf{B}}}\overline{\underline{\mathbf{C}}})$.

A.5 Least Squares Method for Partial Regression Lines

The least squares method tries to find the parameters of a given function class by minimizing the sum of residua between the function curve and the scan data. The weight of large residua should be chosen large in comparison to small residua. The relation between the function points and the N_{scan} data pairs $\left\langle x^{(i)}, z^{(i)} \right\rangle$ can be expressed by the polynomial

$$\mathbf{z} = f(\mathbf{x}) = \sum_{r=1}^{m} a_r \mathbf{x}^{r-1}, \tag{A.5}$$

where $a_1, \ldots, a_m$ are the coefficients of the function (here: $m = 2$). Usually, there are more equations than unknown parameters ($N_{\text{scan}} \gg m$), so that the linear equation system

$$\underbrace{\begin{pmatrix} 1 & x_1 & \cdots & x_1^{m-1} \\ \vdots & & \ddots & \vdots \\ 1 & x_{N_{\text{scan}}} & \cdots & x_{N_{\text{scan}}}^{m-1} \end{pmatrix}}_{\mathbf{M} \in \mathbb{R}^{N_{\text{scan}} \times m}} \cdot \underbrace{\begin{pmatrix} a_1 \\ \vdots \\ a_m \end{pmatrix}}_{a \in \mathbb{R}^m} = \underbrace{\begin{pmatrix} z_1 \\ \vdots \\ z_{N_{\text{scan}}} \end{pmatrix}}_{z \in \mathbb{R}^{N_{\text{scan}}}} \tag{A.6}$$

is inconsistent and the equation $\mathbf{Ma} = \mathbf{z}$ can usually not be solved exactly. Therefore, one computes the solution for $\mathbf{a}$ by minimizing the squared error:

$$\min \left(\|\mathbf{Ma} - \mathbf{z}\|^2 \right) = \min \left(\left(\sum_{r=1}^{m} a_r x_1^{r-1} - z_1 \right)^2 + \ldots + \left(\sum_{r=1}^{m} a_r x_{N_{\text{scan}}}^{r-1} - z_{N_{\text{scan}}} \right)^2 \right), \tag{A.7}$$

which can be rewritten as

$$\min \left(\mathbf{a}^{\mathrm{T}} \mathbf{M}^{\mathrm{T}} \mathbf{Ma} - 2\mathbf{a}^{\mathrm{T}} \mathbf{M}^{\mathrm{T}} \mathbf{z} + \mathbf{z}^{\mathrm{T}} \mathbf{z} \right). \tag{A.8}$$

If one differentiates Equation A.8 with respect to $\mathbf{a}$ and sets the result to zero, one obtains:

$$0 = \mathbf{M}^{\mathrm{T}} \mathbf{Ma} + \mathbf{a}^{\mathrm{T}} \mathbf{M}^{\mathrm{T}} \mathbf{M} - 2\mathbf{M}^{\mathrm{T}} \mathbf{z}, \tag{A.9}$$

$$0 = \mathbf{M}^{\mathrm{T}} \mathbf{Ma} + \mathbf{M}^{\mathrm{T}} \mathbf{z}. \tag{A.10}$$

Consequently, if $\mathbf{M}^{\mathrm{T}} \mathbf{M}$ is invertible, the optimal solution $\mathbf{a}^*$ for $\mathbf{a}$ is

$$\mathbf{a}^* = (\mathbf{M}^{\mathrm{T}} \mathbf{M})^{-1} \mathbf{M}^{\mathrm{T}} \mathbf{z}. \tag{A.11}$$

These parameters provide the coefficients of the partial regression line.

A.6 Flow Chart of the DBSCAN Algorithm

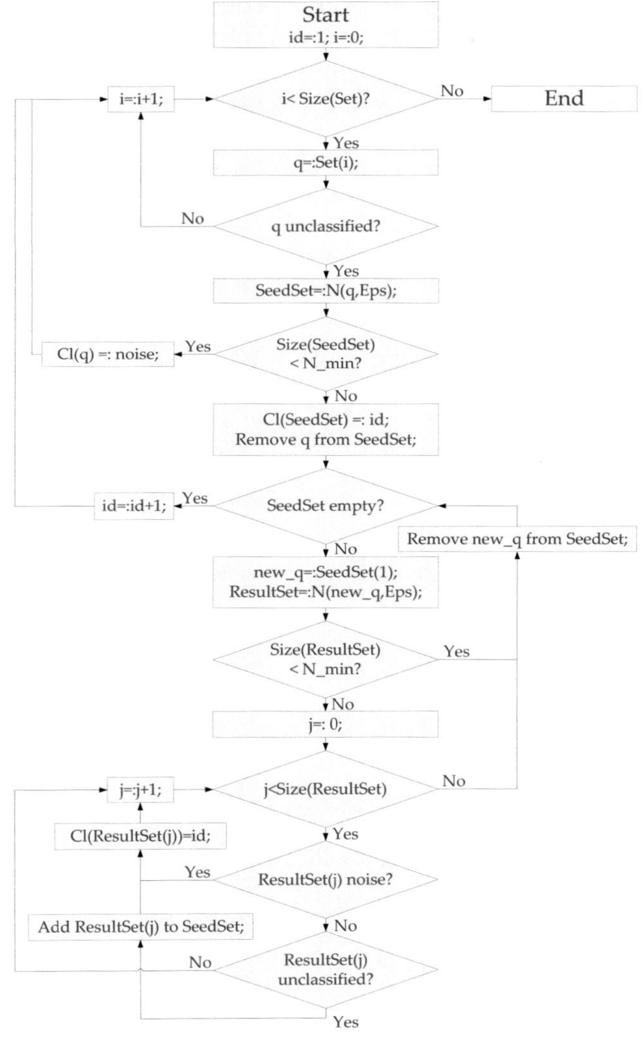

Figure A.1 Flow chart of DBSCAN algorithm: cluster (Cl), ϵ-neighborhood of scan point q (N(q,Eps)), S^{seed} (SeedSet).

A.7 Measurement Results of Short Range Radar

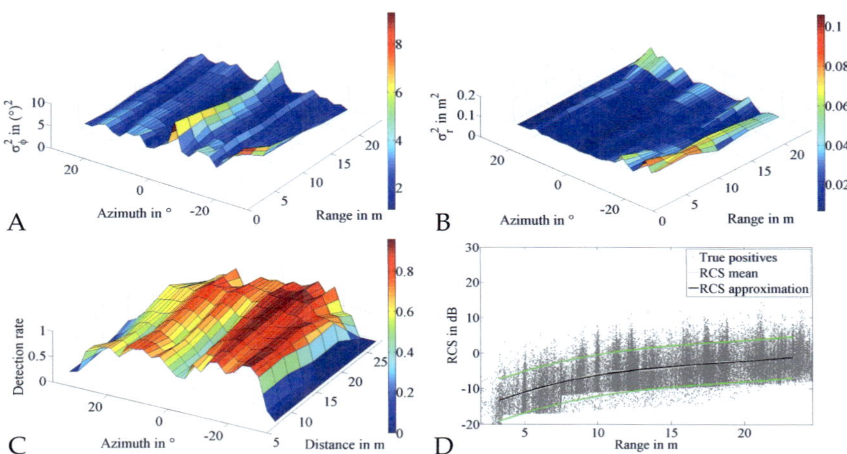

Figure A.2 Determined angular and radial variance and detection rate (C) of the SRR for standing pedestrians: The sensor detects only legs, especially in the near range. Additional variance in the pedestrian position due to walking is not included here, but added later. Subgraph D shows the Radar Cross Section (RCS) values of pedestrians depending on the range.

A.8 Additional Tracking Scenarios

Figure A.3   Examples for different tracking scenarios: Boxes mark tracked objects. The darkness of the box represents the PoE. The lighter the box is, the higher is the PoE. Yellow ellipses represent the 3σ ranges of the track positions. (The legend is provided in the text at the beginning of Subsection 6.1.3.)

Figure A.4 illustrates a scene where two pedestrians follow the lane on the right pavement. There are some uncertain double HOG detections for the pedestrian at the larger distance. The JIPDA sets up tracks that die after a few cycles. In contrast, the EKF01 keeps the track and sets up a new track for the closer detections that follow. This track is then associated with radar detections resulting from the column of the traffic light. A stable EKF01-track is kept on this column which is not a relevant object.

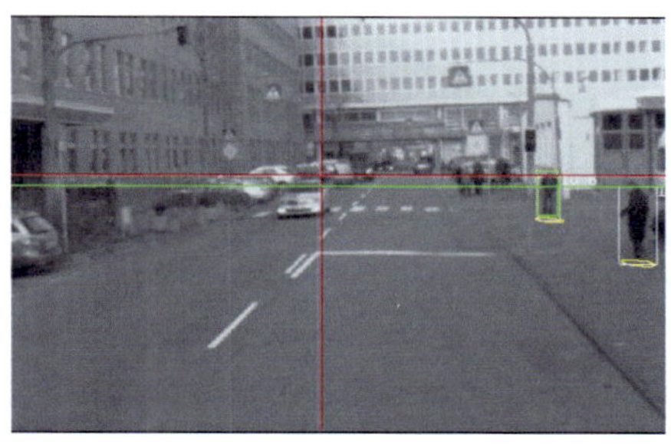

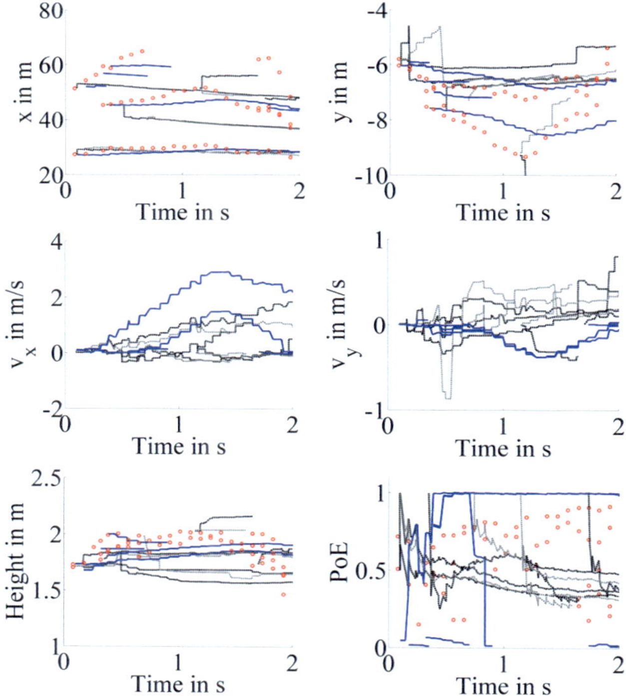

Figure A.4 Scenario with two pedestrians following the lane on the right pavement in different longitudinal distances. One track of the EKF01 is pulled to the column of the traffic light by radar detections.

Abbreviations and Symbols

General Abbreviations

Abbreviation	Meaning
3D	Three-dimensional
ABS	Anti-lock Braking System
ACC	Adaptive Cruise Control
A/D	Analogue to Digital
ADAS	Advanced Driver Assistance System
ANN	Artificial Neural Network
ASIL	Automotive Safety Integrity Level
AUC	Area Under Curve
BBA	Basic Belief Assignment
bps	Bits per Second
BSR	Blind Spot Radar
CAN	Controller Area Network
CDF	Cumulative Distribution Function
cJPDA	Cheap Joint Probabilistic Data Association
CMOS	Complementary Metal Oxide Semi-Conductor
CPHD	Cardinalized Probability Hypothesis Density
CSMA/CA	Carrier Sense Multiple Access / Collision Avoidance
CTRV	Constant Turn Rate and (Constant) Velocity
CVCO	Constant Velocity and Constant Orientation
CW	Continuous Wave (Radar)
DAC	Digital Analogue Converter
DBSCAN	Density-Based Spatial Clustering of Applications with Noise
DMMR	Differential Maximum Minimum Range
DPHD	Dimensional Probabilistic Hypothesis Density
DIN	German Industry Norm (German: Deutsche Industrienorm)
DTW	Dynamic Time Warping
ECU	Electronic Control Unit
EKF	Extended Kalman Filter
ERM	Empirical Risk Minimization
ESC	Electronic Stability Control

Abbreviation	Meaning
DAS	Driver Assistance System
FCW	Forward Collision Warning
FFT	Fast Fourier Transform
FISST	Finite Set Statistics
FMCW	Frequency Modulated Continuous Wave (radar)
FN	False Negative
FOV	Field of View
FP	False Positive
FPS	Frames Per Second
GIDAS	German In-Depth Accident Study
GLM	Generalized Linear Model
GMR	Giant Magneto Resistive
GNN	Global Nearest Neighbor (data association)
GNP	Gross National Product
GPS	Global Positioning System
HDRI	High Dynamic Range Imaging
HMM	Hidden Markov Model
HOG	Histograms of Oriented Gradients
IAR	Insurers Accident Research
ID	Identification Number
IMM	Interacting Multiple Model
IQR	Inter-Quartile Range
IPDA	Probabilistic Data Association with Integrated existence estimation
ISO	International Organization for Standardization
JIPDA	Joint Integrated Probabilistic Data Association
JPDA	Joint Probabilistic Data Association
JPDAM	JPDA with Unresolved Measurements
LASER	Light Amplification by Stimulated Emission of Radiation
LCS	Longest Common Subsequence
LDW	Lane Departure Warning
LIDAR	Light Detection and Ranging
LNN	Local Nearest Neighbor (data association)
LOS	Line of Sight
LRR	Long Range Radar
LVDS	Low Voltage Differential Signaling
MAP	Maximum A-Posteriori
MEMS	Micro-Electro-Mechanical Systems
MHI	Multi-Hypothesis Initialization
MHT	Multi-Hypothesis Tracker
ML	Maximum Likelihood

Abbreviation	Meaning
MLP	Multi-Layer Perceptron
MUX	Multiplexer
MV	Maneuver
NCPS	Normalized Projected Correction Squared
NEES	Normalized Estimation Error Squared
NIR	Near Infrared
NIS	Normalized Innovation Error Squared
NMS	Non-Maximum Suppression
NNSF	Nearest Neighbor Standard Filter
OOSM	Out of Sequence Measurement
PD	Phase Distortion
PDA	Probabilistic Data Association
PDF	Probability Density Function
PHD	Probabilistic Hypothesis Density
PMD	Photonic Mixer Device
px	Pixel
PoE	Probability of Existence
RCS	Radar Cross Section
RF	Radio Frequency
RMS	Root Mean Squared (Error)
ROC	Receiver Operating Characteristic
ROI	Region of Interest
RRT	Rapidly-exploring Random Tree algorithm
SLA	Speed Limit Assist
SMC-MTB	Sequential Monte-Carlo Multi-Target Bayes Filter
SMO	Sequential Minimal Optimization
SNN	Shared Near Neighbor (algorithm)
SNR	Signal-to-Noise Ratio
SRR	Short Range Radar
SUE	Single Use Expression
SVM	Support Vector Machine
TCP/IP	Transmission Control Protocol / Internet Protocol
TN	True Negative
ToF	Time of Flight
TP	True Positive
TTA	Time-To-Appear
TTC	Time-To-Collision
TTD	Time-To-Disappear
UDP	User Datagram Protocol
UKF	Unscented Kalman Filter
USB	Universal Serial Bus
UTC	Universal Time Coordinated

Abbreviation	Meaning
VCO	Voltage Controlled Oscillator
vFSS	Forward Looking Safeguard Systems (German: vorausschauende Frontschutzsysteme)
WLAN	Wireless Local Area Network

General Functions and Operators

Symbol	Meaning
$\dim(\mathbf{x})$	Dimension of vector $\mathbf{x}$
$\text{dist}(\cdot,\cdot)$	Returns a distance measure
$\exp(\cdot)$	Exponential function
$f(\cdot)$	Process function
$f_{\text{kernel}}(\cdot)$	Kernel density function
$f_{\text{flow}}(\cdot)$	Flow function for a specific mode
$f(\delta)$	Probability distribution of lateral dynamics
$f(s)$	Probability distribution of longitudinal dynamics
$f(\xi, t_k)$	Road user's probability distribution at time point t_k
$g(\cdot)$	Controller function
$g_{\text{guard}}(\cdot)$	Mapping function for resulting in a guard set
$h(\cdot)$	Measurement function
$h_{\text{jump}}(\cdot)$	Jump function to the next continuous state after transition
$h_{\Theta}(\cdot)$	Hypothesis function dependent on parameters Θ
$\text{ind}(\cdot)$	Indicator function
$\text{inv}(\cdot)$	Invariant function of a hybrid automaton
$\min(\cdot)$	Returns the smallest value of its content
$\text{square}(\cdot,\cdot)$	Evaluates if arguments $(\cdot,\cdot)$ are within a particular square
$E(\cdot)$	Expectation value
$K(\cdot)$	Kernel distribution function
$N_{\epsilon}(\cdot)$	ϵ-neighborhood
$O(\cdot)$	Complexity
$P(\cdot)$	Operator determining the probability of an event
$\delta_{\text{KD}}(\cdot)$	Kronecker delta function
$\delta(\cdot)$	Probability density function of the Dirac distribution
$\delta_{\mathcal{X}}(\cdot)$	Set-valued Dirac-Delta function
$\Xi(\cdot)$	Returns the coverage of two areas

Variables and Constants

Symbol	Meaning
a	Acceleration
a_{lc}	Coefficient describing the yaw rate of a lane change prototype
a_{max}	Maximum reachable acceleration by a particular class of road users
a_n	Normal acceleration
a_r	Coefficient of a polynomial describing a regression line
a_{start}	Acceleration of pedestrian who starts walking
a_t	Tangential acceleration
a_{turn}	Coefficient to describe yaw rate of turn prototype
b	Kernel bandwidth
b_{lc}	Coefficient describing the yaw rate of a lane change prototype
b_{turn}	Coefficient describing the yaw rate of a turn prototype
c	Speed of light
c_{lc}	Coefficient describing the yaw rate of a lane change prototype
c_{turn}	Coefficient describing the yaw rate of a turn prototype
const	Arbitrary constant value
$\overline{d}$	Path deviation segment
d^{gate}	Gating parameter (threshold) for given gating probability
d_{lc}	Coefficient describing the yaw rate of a lane change prototype
d^{mh}	Mahalanobis distance
d_{tree}	Level of depth in a hypothesis tree
d_{turn}	Coefficient describing the yaw rate of a turn prototype
e	Single association
$e_{L(d)}$	Leave node
f_{opt}	Focal length
$frac_{FN}$	Fraction of false negative detections
$frac_{TP}$	Fraction of true positive detections
$frac_{TP}^{MB}$	Fraction of true positive and maybe-detections
g	Gravity constant
k_g	Gating distance parameter

Symbol	Meaning
l	Length in meters
l_k	Arc element (trajectory classification)
l_{tr}	Length of a trajectory in meters
mv	Maneuver class
n_{lcs}	Normalized longest common subsequence value
o_{rel}	Overlap fraction of A_{label} and A_{track}
p_B	Probability of an object birth
p^{crash}	Crash probability the system vehicle with another road user
p^{dev}	Deviation segment probability of a road user
p_D	Probability of object detection
p_{fasso}	Probability of a false association between an object and a measurement
p_{FP}	Probability of a false positive detection
p_g	Gating probability
p^{int}	Intersection probability of two position trapezoids
p_I	Overlap probability of an object and a hypothesis
p^{occ}	Occlusion probability of an object
p^{path}	Path segment probability of a road user
p^{pos}	Probability of a road user to be located within a certain position trapezoid
p_S	Survival probability
p^{traj}	Probability of a trajectory
p_{TP}	Probability of a true positive detection
$p^{\mathcal{H}}_{\Lambda}$	Likelihood probability of a hypothesis
$p(\exists x_i)$	Existence probability of object x_i
pos_{brake}	Brake pedal position
q	Mode / discrete state (hybrid automaton)
q	Laser scanner point
q^{center}	Center point
q^{core}	Core point of a cluster
q^{seed}	Seed point
r	Range
$\dot{r}$	Range rate
r_{lm}	Elements of the rotation matrix $\mathbf{R}_{rot}$
s_j	Seed point (clustering)
s	Longitudinal position on a path
s	Size of an object in the real world
s^I	Size of an object in the image
$\underline{\bar{s}}$	Path segment
s_{detect}	Maximum detection range of a road user

Symbol	Meaning
t	Continuous time
t_h	Temporal prediction horizon
t^inv	Point of time when reachable sets leaves the invariant
t_k	Time step k
t_start	Acceleration duration of a pedestrian until he reaches his final speed
u	Input / acceleration input (hybrid automaton)
u	Horizontal image coordinate
v	(Tangential) velocity / speed
v	Vertical image coordinate
v_sw	Threshold speed for switching the acceleration model
v_max	Maximum allowed speed for a corresponding class of road users
v_x	Longitudinal speed
v_y	Lateral speed
w	Width of a road user
x	Longitudinal position in Cartesian coordinates or in a curve-aligned coordinate system (front positive)
y	Lateral position in Cartesian coordinates or in a curve-aligned coordinate system (left positive)
z	Vertical position in Cartesian coordinates (up positive)
x	Object in tracking
y	Target variable in classification
z	Measurement in tracking
z	Discrete state of a Markov chain
z^I	System detection in image
A_label	Area of a labeled box representing the existence of a pedestrian
A_track	Area of a box representing the projection of a tracked pedestrian object
$A_\cap$	Intersection area of A_label and A_track
M_u	Horizontal value of optical center
M_v	Vertical value of optical center
V	Volume
α	Yaw angle of a sensor
α_incl	Body inclination angle
β	Pitch angle of a sensor
β_{bj}	Association weight for object birth

Symbol	Meaning
γ	Roll angle of a sensor
δ	Path deviation
ϵ	Threshold distance for the neighborhood of a point (clustering)
ϵ_{N_a}	Minimum number of measurement-object associations in a row for initialization
ϵ_{NEES}	Normalized estimation error squared
$\bar{\epsilon}_{NEES}$	Mean of the normalized estimation error squared based on all objects and time points
ϵ_p	Threshold probability for a predicted state distribution (cancellation)
$\hat{e}_{\mathcal{S}_j}(h_{\Theta,ij})$	Cross validation error
$\epsilon_{\dot{\phi}_e}$	Threshold value for an evanescent yaw rate
λ_{tf}	Wavelength of a transmitted signal
η	Normalization constant
σ	Standard deviation
τ_d	Dispersion parameter
τ_t	Travel time or time of flight
τ_r^{wc}	Worst case runtime
τ	Time interval of prediction in reachability analysis
θ_k	Angle between two line segments
ϕ	Yaw angle of the system vehicle
$\dot{\phi}$	Yaw rate of the system vehicle
ϕ_{steer}	Steering wheel angle
$\tilde{\zeta}_{del}$	Threshold value for object deletion
$\tilde{\zeta}_{PoE}$	Threshold value for object confirmation
ζ_B	Threshold value for object instantiation
Δt	Time interval
Δt_d	Duration of a trajectory
Λ	Likelihood
Θ	Parameters of the world (hypothesis functions)

Matrices and Vectors

Symbol	Meaning
$\mathbf{l}_k$	Line segment between two successive measurement points (k and $k+1$)
$\mathbf{p}$	Probability vector
$\tilde{\mathbf{p}}$	Combined probability vector of state and input
$\mathbf{t}$	Translation vector

Symbol	Meaning
$\mathbf{u}$	Input vector
$\mathbf{x}$	State vector and feature vector in classification
$\mathbf{x}^{\mathcal{H}}$	State vector of an object hypothesis
$\mathbf{x}_{\text{ref}}$	Reference state vector
$\mathbf{z}$	Measurement vector
$\mathbf{z}_{\text{dev}}$	Measurement deviation vector
$\mathbf{z}_{\text{ref}}$	Reference measurement vector
$\mathbf{z}_{\text{S}}$	Measurement vector of measured sensor
$\mathbf{A}$	Association matrix
$\mathbf{A}_{\text{global}}$	Globally optimal, binary association matrix
$\mathbf{C}$	Cost matrix in classification
$\mathbf{C}$	Matrix for longest common subsequence computation
$\mathbf{D}$	Distance matrix
$\mathbf{E}$	Matrix with classification fractions
$\mathbf{F}$	System matrix
$\mathbf{G}$	Gating matrix
$\mathbf{H}$	Measurement matrix
$\mathbf{K}$	Kalman gain
$\mathbf{M}$	Matrix containing laser scanner data of one dimension
$\mathbf{P}$	State covariance matrix
$\mathbf{Q}$	Process noise covariance matrix
$\mathbf{R}$	Measurement noise covariance matrix
$\mathbf{R}_{\text{rot}}$	Rotation matrix with elements r_{lm}
$\mathbf{S}$	Innovation covariance matrix
$\mathbf{T}$	Transformation matrix
$\mathbf{T}_{(\cdot,\cdot)}$	Trajectory
$\boldsymbol{\beta}$	Weighting matrix for probabilistic data association
$\boldsymbol{\gamma}$	Measurement residuum
$\boldsymbol{\gamma}_{\text{ref}}$	Deviation vector of the estimated state and the reference state (ground truth)
$\boldsymbol{\epsilon}$	Hypothesis error in state covariance matrix (PDA)
$\boldsymbol{\epsilon}_{\text{T}}$	Sequence of maximum differences between trajectory symbols for longest common subsequence computation
$\boldsymbol{\rho}$	Parameter vector
$\boldsymbol{\Gamma}$	Noise gain
$\boldsymbol{\Gamma}$	Input transition matrix
$\tilde{\boldsymbol{\Gamma}}$	Input transition matrix in rearranged form to consider uncertain input
$\boldsymbol{\Psi}$	State transition matrix

Symbol	Meaning
$\tilde{\Psi}$	State transition matrix in rearranged form to consider uncertain input

Sets

Symbol	Meaning
$\mathbb{N}$	Set of natural numbers
$\mathbb{N}^+$	Set of positive natural numbers
$\mathbb{R}$	Set of real valued numbers
$\mathbb{R}^+$	Set of positive real-valued numbers
$\mathcal{A}$	Set of two-dimensional positions
$\mathcal{B}$	Set of positions occupied by road user's body
$\mathcal{C}$	Set of positions occupied by road user's center
$\mathcal{C}_{\mathrm{cl}}$	Cluster of objects or of laser scanner points
$\mathcal{D}_{\mathrm{points}}$	Dataset of laser scanner points
$\mathcal{D}$	Probabilistic hypothesis density function
$\mathcal{D}^{\mathrm{d}}$	Dimensional probabilistic hypothesis density function
$\tilde{\mathcal{D}}^{\mathrm{d}}$	Discrete DPHD function
$\mathcal{E}_d$	Single association hypothesis (set of single assignments)
$\mathcal{E}_{ij}^{\mathrm{TP}}$	Set of hypotheses stating a true positive association of object x_i
$\mathcal{E}_i^{\exists}$	Set of hypotheses stating existence of object x_i
$\mathcal{G}$	Guard set
$\mathcal{H}$	Class of hypothesis functions
$\mathcal{I}$	Set of real valued intervals
$\mathcal{M}$	Set of models for classification
$\mathcal{N}(\mu, \sigma)$	Normal distribution with mean μ and standard deviation σ
$\mathcal{P}$	Set of parameters
$\mathcal{R}$	Reachable set
$\mathcal{R}^+$	Reachable set after transition to a new mode
$\mathcal{R}^{\mathrm{int}}$	Intersection of reachable set with a hit guard set
$\mathcal{S}$	Set of training examples in classification
$\mathcal{S}_{\mathrm{train}}$	Training data set for cross-validation
$\mathcal{S}_{\mathrm{cv}}$	Cross validation set for cross-validation
$\mathcal{S}^{\mathrm{noise}}$	Set of points representing noise
$\mathcal{S}_q$	Laser scanner point set
$\mathcal{S}^{\mathrm{seed}}$	Set of seed points

Symbol	Meaning
$\mathcal{T}$	Set of discrete transitions of a hybrid automaton
$\mathcal{U}$	Set of inputs
$\mathcal{X}$	Set of states
$\mathcal{X}_\mathrm{a}$	Set of objects assigned to a measurement
$\mathcal{X}_\mathrm{env}$	Set of object in the environment model
$\mathcal{X}_\mathrm{na}$	Set of objects that could not be assigned to a measurement
$\mathcal{X}^*$	Set of object indexes and corresponding special symbols
$\mathcal{Z}^*$	Set of measurement indexes and corresponding special symbols
ν	Stochastic process noise distribution
η	Stochastic measurement noise distribution
Ω	Set of elementary events

Quantities

Symbol	Meaning
$\tilde{n}$	Number of intermediate points of time for Markov chain abstraction
n	Number of state dimensions
n_ϵ	Number of points within radius ϵ around a point
m	Degree of a polynomial
M	Number of measurements per time step
M^c	Number of measurements in a cluster
M^g	Number of measurements within a gating region of an object
$\hat{M}^\mathrm{fa}$	Expectation value for the number of false alarms
M_lc	Number of symbols in a prototype trajectory for a lane change
M_lf	Number of symbols in a prototype trajectory for lane following
M_p	Number of symbols in a prototype trajectory
M_turn	Number of symbols in a prototype trajectory for turn
N	Number of objects in the environment model
N^*	Cardinality
N_a	Number of measurement associations since object birth
N_asso	Number of detections associated with references

Symbol	Meaning
N^c	Number of objects in a cluster
N_{cell}	Number of measurement associations in a grid cell
N_{comb}	Number of possible combinations
N_{cl}	Number of clusters
N_{class}	Number of classes in a dataset
N_{cont}	Number of continuous states
N_{det}	Number of detections
N_{dis}	Number of discrete states
N_ϵ	Number of points in the neighborhood of center point q^{center} within radius ϵ
N_f	Number of features in classification
N_{free}	Degree of freedom
N_{it}	Number of iterations
N_l	Number of lines
N_{min}	Minimum number of points in the neighborhood of a point (clustering)
N_{mode}	Number of modes (discrete states)
N_{na}	Number of missing measurement associations since object birth
N_{node}	Number of nodes
N_p	Number of parameters
N_{points}	Number of considered points
N_{ref}	Number of references
N_s	Number of subsets in cross-validation
N^{sample}	Number of available samples for kernel density estimation
N_{scan}	Number of scan points per cycle
N_{sen}	Number of sensor models
N^{sim}	Number of simulation runs for Markov chain abstraction
N_t	Number of training examples in classification
N_{tol}	Maximum number of symbols that may be between similar symbols
N_{tr}	Number of samples building a trajectory
N_u	Number of input cells for Markov chain abstraction
N_x	Number of state cells for Markov chain abstraction
$N_\mathcal{E}$	Number of association hypotheses
N_{FN}	Number of false negative detections
N_{TP}	Number of true positive detections
N_{TP}^{MB}	Number of true positive and true positive maybe-detections

Subscripts and Superscripts

Symbol	Meaning
$\Box_d$	Association hypothesis index
$\Box^{\text{dev}}$	Relates to the deviation of a road user from the path
$\Box_e$	Path segment index of the system vehicle
$\Box_\text{e}$	Relates to the system vehicle
$\Box_f$	Deviation segment index of the system vehicle
$\Box_g$	Path segment index of a road user
$\Box_h$	Deviation segment index of a road user
$\Box_i$	(Object or cell) index
$\Box_j$	(Measurement or cell) index
$\Box_k$	Time step / time point index
$\Box_l$	Velocity segment index or dimension index
$\Box_\text{lc}$	Refers to maneuver class lane change
$\Box_\text{lf}$	Refers to maneuver class lane following
$\Box_m$	Input dimension
$\Box_n$	State dimension
$\Box_p$	Parameter vector dimension
$\Box_\text{p}$	Refers to a prototype trajectory
$\Box^{\text{path}}$	Relates to a position on the path of a road user
$\Box_\text{ped}$	Relates to a pedestrian
$\Box^{\text{pos}}$	Relates to the two-dimensional position of a road user
$\Box_\text{ref}$	Relates to reference measurements (ground truth)
$\Box_\text{ru}$	Relates to another road user
$\Box_s$	Scan point index
$\Box_t$	Scan point index
$\Box_\text{turn}$	Refers to the maneuver class turn
$\Box_x$	Longitudinal component of $\Box$
$\Box_y$	Lateral component of $\Box$
$\Box_z$	Vertical component of $\Box$
$\Box^\text{C}$	Relates to a camera
$\Box_\text{CC2VC}$	Camera to vehicle coordinates
$\Box_\text{H}$	Homogeneous coordinates
$\Box^\text{R}$	Relates to a radar
$\Box_\text{RC2VC}$	Radar to vehicle coordinates
$\Box^\text{S}$	Relates to any sensor
$\Box_\text{SC2VC}$	Sensor to vehicle coordinates
$\Box_\text{VC}$	Relates to vehicle coordinates
$\Box^\alpha$	Input index

Symbol	Meaning
$\square^{\beta}$	Input index

General Spellings

Symbol	Meaning
$\left\langle x^{(i)}, y^{(i)} \right\rangle$	Training example in classification
$\left\langle x^{(i)}, z^{(i)} \right\rangle$	Laser scanner data pair
$\left\langle q_i, q_j \right\rangle$	Laser scanner point pair
$\underline{\overline{\square}}$	Interval set
$\overline{\square}$	Upper interval limit of $\underline{\overline{\square}}$
$\underline{\square}$	Lower interval limit of $\underline{\overline{\square}}$
$\hat{\square}$	Estimated mean of variable $\square$

Bibliography

[1] DEKRA Automobil GmbH, "DEKRA Verkehrssicherheitsreport Lkw," 2009.

[2] ——, "DEKRA Verkehrssicherheitsreport – Keine Entwarnung für Radfahrer und Fußgänger," 2011.

[3] World Health Organization, *World report on road traffic injury prevention: Summary*, vol. 4, 2004.

[4] Volvo Car Corporation, "Volvo S60 – Five things to know." [Online]. Available: http://www.volvocars.com/intl/all-cars/volvo-s60/pages/5-things.aspx (last visit: February 2nd, 2012), 2012.

[5] C. Stiller and M. Maurer, Eds., *Fahrerassistenzsysteme mit maschineller Wahrnehmung*. Springer, Berlin Heidelberg New York, 2005.

[6] Daimler AG, "25 Jahre ABS für Nutzfahrzeuge: Start in die moderne Sicherheitstechnik." [Online]. Available: http://media.daimler.com/dcmedia/0-921-657345-49-820701-1-0-0-0-0-0-11701-614240-0-1-0-0-0-0-0.html (last visit: November 27th, 2012), 2006.

[7] UNESCO, "Vienna convention on road traffic," in *United Nations Economic and Social Council's Conference*, November 1968.

[8] Deutsche Presse-Agentur GmbH, "Hoher Dieselpreis belastet Unternehmen." [Online]. Available: http://www.handelsblatt.com/unternehmen/industrie/transport-branche-hoher-dieselpreis-belastet-unternehmen/7036548.html (last visit: November 27th, 2012), 2012.

[9] ISO TC22/SC3/WG16, "ISO/DIS 26262 – Road vehicles – Functional safety (Draft International Standard)," 2009.

[10] M. Skolnick, *Introduction to Radar Systems*. McGraw-Hill, 1980.

[11] A.D.C. GmbH / Continental, "Data ARS300." [Online]. Available: http://www.conti-online.com (last visit: January 9th, 2011), 2011.

[12] Velodyne Lidar, "Datasheet: High definition lidar HDL-64E S2." [Online]. Available: http://velodynelidar.com/lidar/products/brochure/HDL-64E (last visit: November 27th, 2012), 2010.

[13] L. Wald, "Some terms of reference in data fusion," *IEEE Transactions on Geosciences and Remote Sensing*, vol. 37, no. 3, pp. 1190–1193, 1999.

[14] J. Beyerer, F. Puente León, and K.-D. Sommer, Eds., *Informationsfusion in der Mess- und Sensortechnik*. Universitätsverlag Karlsruhe, Karlsruhe, 2006.

[15] H. Durrant-Whyte, "Sensor model and multisensory integration," *International Journal of Robotics Research*, vol. 6, pp. 3–24, 1988.

[16] F. Puente León and H. Ruser, "Information fusion – Overview and taxonomy," *Reports on Industrial Information Technology*, vol. 12, pp. 1–18, 2010.

[17] B. Ristic, S. Arulampalam, and N. Gordon, *Beyond the Kalman Filter: Particle Filters for Tracking Applications*. Artech House Publishers, 2004.

[18] R. E. Kalman, "A new approach to linear filtering and prediction problems," *Transactions of the ASME – Journal of Basic Engineering*, vol. 82, no. Series D, pp. 35–45, 1960.

[19] G. Welch and G. Bishop, "An introduction to the Kalman filter," *In SIGGRAPH Course Notes*, 2001.

[20] S. J. Julier and J. K. Uhlmann, "New extension of the Kalman filter to nonlinear systems," *Signal Processing, Sensor Fusion, and Target Recognition*, vol. 3068, pp. 182–193, 1997.

[21] S. Julier, J. Uhlmann, and H. Durrant-Whyte, "A new method for the nonlinear transformation of means and covariances in filters and estimators," *IEEE Transactions on Automatic Control*, vol. 45, no. 3, pp. 477–482, March 2000.

[22] S. Julier and J. Uhlmann, "Unscented filtering and nonlinear estimation," *Proceedings of the IEEE*, vol. 92, no. 3, pp. 401–422, March 2004.

[23] Y. Bar-Shalom and X.-R. Li, *Estimation and Tracking – Principles, Techniques and Software*. Artech House, 1993.

[24] A. Mutambara, *Decentralized Estimation and Control for Multisensor Systems*. CRC Press, 1998.

[25] J. Beyerer, *Verfahren zur quantitativen statistischen Bewertung von Zusatzwissen in der Meßtechnik*. VDI Verlag, Düsseldorf, 1999.

[26] Shafer and Glenn, *A mathematical Theory of Evidence*. Princeton University Press, 1976.

[27] X. Gros, *NDT Data Fusion*. Arnold, London, 1997.

[28] T. Henderson and H. F. Durrant-Whyte, "Multisensor data fusion," *Handbook of Robotics*, 2008.

[29] L. C. Jain, A. Filippidis, and N. Martin, "Fusion of intelligent agents for the detection of aircraft in SAR images," *IEEE Transactions on Pattern Analysis and Machine Intelligence*, vol. 22, no. 4, pp. 378–384, 2000.

[30] ISO 31000, "ISO 31000 – Guidelines on principles and implementation of risk management," 2009.

[31] V. Popovich, A. Prokaev, O. Smirnova, and F. Galiano, "Combined approach to tactical situations analysis," in *Proc. of the IEEE Military Communications Conference*, pp. 1–7, November 2008.

[32] V. L. Neale, T. A. Dingus, J. Klauer, J. Sudweeks, and M. Goodman, "An overview of the 100-car naturalistic study and findings," 2005.

[33] E. Brühning, D. Otte, and C. Pastor, "30 Jahre wissenschaftliche Erhebungen am Unfallort für mehr Verkehrssicherheit," *Zeitschrift für Verkehrssicherheit*, vol. 51, pp. 175–181, 2005.

[34] H. Brunner and A. Georgi, "Drei Jahre Verkehrsunfallforschung an der TU Dresden," *Automobiltechnische Zeitschrift*, February 2003.

[35] T. Hummel, M. Kühn, J. Bende, and A. Lang, "Fahrerassistenzsysteme – Ermittlung des Sicherheitspotenzials auf Basis des Schadengeschehens der Deutschen Versicherer," 2011.

[36] T. Carthy, D. Packham, D. Salter, and D. Silcock, *Risk and safety on the roads: The older pedestrian*. Foundation for Road Safety Research, The University of Newcastle upon Tyne, 1995.

[37] J. A. Oxley, E. Ihsen, B. N. Fildes, J. L. Charlton, and R. H. Dayc, "Crossing roads safely: An experimental study of age differences in gap selection by pedestrians," *Accident Analysis and Prevention*, vol. 37, pp. 962–971, 2005.

[38] V. Rafaely, J. Meyer, I. Zilberman-Sandler, and S. Viener, "Perception of traffic risks for older and younger adults," *Accident Analysis & Prevention*, vol. 38, no. 6, pp. 1231–1236, 2006.

[39] J. A. Langlois, P. M. Keyl, J. Guralnik, D. J. Foley, R. A. Marottoli, and B. Wallace, "Characteristics of older pedestrians who have difficulty crossing the road," *American Journal of Public Health*, vol. 87, no. 3, pp. 393–397, 1997.

[40] I. M. Bernhoft and G. Carstensen, "Preferences and behaviour of pedestrians and cyclists by age and gender," *Transportation Research Part F*, vol. 11, pp. 83–95, 2008.

[41] W. Niehwoehner and J. Gwehenberger, "Advanced forward-looking safety systems: Pedestrian accident analysis." [Online]. Available: http://www.carhs.de/newsletter-archive/safetynews-2010-12-01-en.html (last visit: January 6th, 2011), 2010.

[42] V. Himanen and R. Kumala, "An application of logit models in analysing the behaviour of pedestrians and car drivers on pedestrian crossings," *Accident Analysis and Prevention*, vol. 20, no. 3, pp. 187–197, 1988.

[43] S. Das, C. F. Manski, and M. D. Manuszak, "Walk or wait? an empirical analysis of street crossing decisions," *Journal of Applied Econometrics*, vol. 20, no. 4, pp. 529–548, 2003.

[44] S. Schmidt and B. Färber, "Pedestrians at the kerb – Recognising the action intentions of humans," *Transportation Research Part F*, vol. 12, pp. 300–310, 2009.

[45] N. Carey, "Establishing pedestrian walking speeds," *Portland State University ITE Student Chapter*, 2005.

[46] K. Teknomo, *Microscopic Pedestrian Flow Characteristics: Development of an Image Processing Data Collection and Simulation Model.* PhD-Thesis, Tohoku University Japan, 2002.

[47] R. R. Knoblauch, M. Pietrucha, and M. Nitzburg, "Field studies of pedestrian walking; speed and start-up time," *Transportation Research Record, Pedestrian and Bicycle Research*, vol. 1538, pp. 27–38, 1996.

[48] F. Kramer and M. Raddatz, "Das Bewegungsverhalten von fußgängern im Straßenverkehr auf Basis einer experimentellen Reihenuntersuchung," *AT-Zonline: Verkehrsunfall und Fahrzeugtechnik*, vol. 12, pp. 328–388, 2010.

[49] T. Bayes, "An essay towards solving a problem in the doctrine of chances," *Philosophical Transactions*, vol. 53, pp. 370–418, 1763.

[50] S. Fienberg, "When did the Bayesian inference become Bayesian," *Bayesian Analysis*, vol. 1, no. 1, pp. 1–40, 2006.

[51] E. Behrends, *Introduction to Markov Chains.* Vieweg, 2000.

[52] R. Mahler, *Statistical Multisource-Multitarget Information Fusion.* Artech House, Inc. Norwood, MA, USA, 2003.

[53] ——, "Multitarget Bayes filtering via first-order multitarget moments," *IEEE Transactions on Aerospace and Electronic Systems*, vol. 39, no. 4, pp. 1152–1178, 2003.

[54] ——, "PHD filters of second order in target number," *Proceedings of SPIE*, vol. 6236, pp. 217–228, 2006.

[55] H. Sidenbladh and S. Wirkander, "Tracking random sets of vehicles in terrain," *Proc. of the Computer Vision and Pattern Recognition Workshop*, vol. 9, 2003.

[56] B. Vo, S. Singh, and A. Doucet, "Random finite sets and sequential Monte Carlo methods in multi-target tracking," *Proc. of the International Radar Conference*, pp. 486–491, 2003.

[57] B. Vo and W. Ma, "The Gaussian mixture probability hypothesis density filter," *IEEE Transactions on Signal Processing*, vol. 54, no. 11, pp. 4091–4104, 2006.

[58] Y. Bar-Shalom and T. E. Fortmann, *Tracking and Data Association*, 179th ed. Academic Press, Inc., Mathematics in Science and Engineering, 1988.

[59] U. Schöning, *Algorithmik.* Spektrum, Akademischer Verlag, 2001.

[60] M. Mählisch, *Filtersynthese zur simultanen Minimierung von Existenz-, Assoziations-, und Zustandsunsicherheiten in der Fahrzeugumfelderfassung mit heterogenen Sensordaten.* PhD Thesis, Universität Ulm, Institut für Mess-, Regel- und Mikrotechnik, 2009.

[61] J. Stirling, "Methodus differentialis," London, 1730.

[62] S. Blackman and R. Popoli, *Design and Analysis of Modern Tracking Systems.* Artech House, Boston, 1999.

[63] A. Frank, "On Kuhn's Hungarian method – A tribute from Hungary / Egervary," *Research report, Research Group on Combinatorial Optimization,* October 2004.

[64] M. Buehren, "Functions for the rectangular assignment problem." [Online]. Available: http://www.mathworks.com/matlabcentral/file-exchange/6543-functions-for-the-rectangular-assignment-problem (last visit: September, 2012), 2004.

[65] M. Munz, *Generisches Sensorfusionsframework zur gleichzeitigen Zustands- und Existenzschätzung für die Fahrzeugumfelderkennung.* PhD Thesis, Universität Ulm, Institut für Mess-, Regel- und Mikrotechnik, 2011.

[66] T. Fortmann, Y. Bar-Shalom, and M. Scheffe, "Sonar tracking of multiple targets using joint probabilistic data association," *IEEE Journal of Oceanic Engineering,* vol. 8, no. 3, pp. 173–184, 1983.

[67] M. Mählisch, M. Szczot, O. Löhlein, M. Munz, and K. Dietmayer, "Simultaneous processing of multitarget state measurements and object individual sensory existence evidence with the joint integrated probabilistic data association filter," in *Proc. of the 5th International Workshop on Intelligent Transportation,* Hamburg, Germany, March 2008.

[68] R. O. Duda, P. E. Hart, and D. G. Stork, *Pattern Classification,* 2nd ed. Wiley, October 2000.

[69] A. Ng and M. I. Jordan, *On Discriminative vs. Generative Classifiers: A comparison of Logistic Regression and Naive Bayes, Neural Information Processing Systems.* iTunes University, Stanford University, 2002.

[70] C. G. Cassandras and S. Lafortune, *Introduction to discrete event systems.* Springer, 2010.

[71] M. Althoff, *Reachability Analysis and its Application to the Safety Assessment of Autonomous Cars.* PhD Thesis, Technische Universität München, 2010.

[72] J. Schröder, *Modelling, State Observation and Diagnosis of Quantized Systems, Lecture Notes in Control and Information Sciences 282.* Springer, Berlin Heidelberg, 2003.

[73] L. Billera, R. Ehrenborg, and M. Readdy, "The cd-index of zonotopes and arrangements," *Mathematical Essays in Honor of Gian-Carlo Rota,* pp. 23–40, 1998.

[74] C. Gardiner, *Handbook of Stochastic Methods: For Physics, Chemistry, and Natural Sciences.* Springer, 1983.

[75] B. Oksendal, *Stochastic Differential Equations.* Springer, 2003.

[76] R. Yuster and U. Zwick, "Fast sparse matrix multiplcation," *ACM Transactions on Algorithms*, vol. 1, pp. 2–13, 2005.

[77] D. Linzmeyer, *Real-Time Detection of Pedestrians from a Moving Vehicle Using Thermopile and Radar Sensor Fusion.* PhD Thesis, Universität Ulm, Fakultät für Ingenieurwissenschaften, 2006.

[78] F. Fayad and V. Cherfaoui, "Object-level fusion and confidence management in a multi-sensor pedestrian tracking system," in *Proc. of the IEEE International Conference on Multisensor Fusion and Integration for Intelligent Systems*, pp. 58–63, August 2008.

[79] R. Sittler, "An optimal data association problem in surveillance theory," *IEEE Transactions on Military Electronics*, vol. 8, no. 2, pp. 125–139, April 1964.

[80] M. Meuter, U. Iurgel, S.-B. Park, and A. Kummert, "The unscented Kalman filter for pedestrian tracking from a moving host," in *Proc. of the IEEE Intelligent Vehicles Symposium*, pp. 37–42, June 2008.

[81] G. Gate, A. Breheret, and F. Nashashibi, "Fast pedestrian detection in dense environment with a laser scanner and a camera," in *Proc. of the IEEE 69th Vehicular Technology Conference*, pp. 1–6, April 2009.

[82] R. O. Chavez-Garcia, J. Burlet, T.-D. Vu, and O. Aycard, "Frontal object perception using radar and mono-vision," in *Proc. of the IEEE Intelligent Vehicles Symposium*, pp. 159–164, 2012.

[83] D. Westenhofen, C. Gründler, K. Doll, U. Brunsmann, and S. Zecha, "Transponder- and camera-based advanced driver assistance system," in *Proc. of the IEEE Intelligent Vehicles Symposium*, pp. 293–298, June 2012.

[84] L. Lamard, R. Chapuis, and J.-P. Boyer, "Dealing with occlusions with multi targets tracking algorithms for the real road context," in *Proc. of the IEEE Intelligent Vehicles Symposium*, pp. 371–376, June 2012.

[85] S. Reuter and K. Dietmayer, "Pedestrian tracking using random finite sets," in *Proc. of the 14th International Conference on Information Fusion*, pp. 1–8, July 2011.

[86] D. Musicki and R. Evans, "Joint integrated probabilistic data association – JIPDA," in *Proc. of the 5th International Conference on Information Fusion*, pp. 1120–1125, 2002.

[87] ——, "Joint integrated probabilistic data association: JIPDA," *IEEE Transactions on Aerospace and Electronic Systems*, vol. 3, pp. 1093–1099, 2004.

[88] Y. Bar-Shalom and E. Tse, "Tracking in a cluttered environment with probabilistic data association," *Automatica*, vol. 11, no. 5, pp. 451–460, 1975.

[89] I. J. Cox, "A review of statistical data association techniques for motion correspondence," *International Journal of Computer Vision*, vol. 10, no. 1, pp. 53–66, 1993.

[90] Y. Bar-Shalom, *Multitarget-Multisensor Tracking: Applications and Advances*, vol. 3. Artech House, Norwood (MA), 2000.

[91] R. J. Fitzgerald, "Development of practical PDA logic for multitarget tracking by microprocessor," in *Proc. of the American Control Conference*, pp. 889–898, 1986.

[92] C. Hoffmann and T. Dang, "Cheap joint probabilistic data association filters in an interacting multiple model design," in *Proc. of the IEEE International Conference on Multisensor Fusion and Integration for Intelligent Systems*, pp. 197–202, September 2006.

[93] J. Roecker and G. Phillis, "Suboptimal joint probabilistic data association," *IEEE Transactions on Aerospace and Electronic Systems*, vol. 29, no. 2, pp. 510–517, April 1993.

[94] P. Horridge and S. Maskell, "Real-time tracking of hundreds of targets with efficient exact JPDAF implementation," in *Proc. of the 9th International Conference on Information Fusion*, pp. 1–8, July 2006.

[95] S. Maskell, M. Briers, and R. Wright, "Fast mutual exclusion," in *Proc. of the SPIE Conference on Signal Processing of Small Targets*, 2004.

[96] S. Zuther, M. Biggel, M. Muntzinger, and K. Dietmayer, "Multi-target tracking for merged measurements of automotive narrow-band radar sensors," in *Proc. of the 12th International IEEE Conference on Intelligent Transportation Systems*, 2009.

[97] V. Kettnaker and R. Zabih, "Bayesian multi-camera surveillance," in *Proc. of the IEEE Conference on Computer Vision and Pattern Recognition*, pp. 2253–2259, June 1999.

[98] T. J. Ellis, D. Makris, and J. K. Black, "Learning a multi-camera topology," in *Proc. of the Joint IEEE International Workshop on Visual Surveillance and Performance Evaluation of Tracking and Surveillance*, pp. 165–171, October 2003.

[99] D. Makris, T. Ellis, and J. Black, "Bridging the gaps between cameras," in *Proc. of the IEEE Conference on Computer Vision and Pattern Recognition*, pp. 205–210, June 2004.

[100] A. Chilgunde, P. Kumar, S. Ranganath, and H. Wei Min, "Multi-camera target tracking in blind regions of cameras with non-overlapping fields of view," in *Proc. of the Bristish Machine Vision Conference*, pp. 397–406, September 2004.

[101] W. Du and J. Piater, "Multi-camera people tracking by collaborative particle filters and principal axis-based integration," *ACCV 2007, Part I, Lecture Notes in Computer Science*, vol. 4843, pp. 365–374, 2007.

[102] J. Black, D. Makris, and T. Ellis, "Validation of blind region learning and tracking," in *Proc. of the 2nd Joint IEEE International Workshop on Visual Surveillance and Performance Evaluation of Tracking and Surveillance*, pp. 9–16, October 2005.

[103] O. Javed, Z. Rasheed, K. Shafique, and M. Shah, "Tracking across multiple cameras with disjoint views," in *Proc. of the IEEE International Conference on Computer Vision*, pp. 952–957, October 2003.

[104] O. Javed, K. Shafique, Z. Rasheed, and M. Shah, "Modeling inter-camera space-time and appearance relationships for tracking across non-overlapping views," *Computer Vision and Image Understanding*, vol. 109, pp. 146–162, 2008.

[105] Y. Wang, L. He, and S. Velipasalar, "Real-time distributed tracking with non-overlapping cameras," in *Proc. of the IEEE 17th International Conference on Image Processing*, September 2010.

[106] Y. R. Loke, P. Kumar, S. Ranganath, and W. M. Huang, "Object matching across multiple non-overlapping fields of view using fuzzy logic," *Acta Automatica Sinica*, vol. 32, no. 36, pp. 978–987, November 2006.

[107] P. L. Mazzeo, P. Spagnolo, and T. D'Orazio, "Object tracking by non-overlapping distributed camera network," *Advanced Concepts for Intelligent Vision Systems, Lecture Notes in Computer Science*, vol. 5807, pp. 516–527, 2009.

[108] F. Porikli and A. Divakaran, "Multi-camera calibration, object tracking and query generation," in *Proc. of the International Conference on Multimedia and Expo*, pp. 653–656, July 2003.

[109] P. KraewTraKulPong and R. Bowden, "Probabilistic learning of salient patterns across spatially separated, uncalibrated views," in *Proc. of the IEEE Intelligent Distributed Surveillance Systems Conference*, pp. 36–40, February 2004.

[110] A. Rahimi, B. Dunagan, and T. Darrell, "Simultaneous calibration and tracking with a network of non-overlapping sensors," in *Proc. of the IEEE Conference on Computer Vision and Pattern Recognition*, vol. 1, pp. I–187–I–194, June 2004.

[111] C. Rabe, "Detection of moving objects by spatio-temporal motion analysis," Ph.D. dissertation, University of Kiel, Kiel, Germany, 2011.

[112] T. Gandhi and M. M. Trivedi, "Pedestrian collision avoidance systems: A survey of computer vision based recent studies," in *Proc. of the IEEE Intelligent Transportation Systems Conference*, pp. 976–981, September 2006.

[113] M. Enzweiler and D. Gavrila, "Monocular pedestrian detection: Survey and experiments," *IEEE Transactions on Pattern Analysis and Machine Intelligence*, vol. 31, no. 12, pp. 2179–2195, December 2009.

[114] ——, "Integrated pedestrian classification and orientation estimation," in *Proc. of the IEEE Conference on Computer Vision and Pattern Recognition*, pp. 982–989, June 2010.

[115] ——, "A multi-level mixture-of-experts framework for pedestrian classification," *IEEE Transactions on Image Processing*, vol. 20, no. 10, pp. 2967–2979, October 2011.

[116] K. Goto, K. Kidono, Y. Kimura, and T. Naito, "Pedestrian detection and direction estimation by cascade detector with multi-classifiers utilizing feature interaction descriptor," in *Proc. of the IEEE Intelligent Vehicles Symposium*, pp. 224–229, June 2011.

[117] N. Dalal and B. Triggs, "Histograms of oriented gradients for human detection," in *Proc. of the International Conference on Computer Vision and Pattern Recognition*, 2005.

[118] C. Keller, M. Enzweiler, and D. M. Gavrila, "A new benchmark for stereo-based pedestrian detection," in *Proc. of the IEEE Intelligent Vehicles Symposium*, June 2011.

[119] C. Wojek, S. Walk, and B. Schiele, "Multi-cue onboard pedestrian detection," in *Proc. of the Conference on Computer Vision and Pattern Recognition*, pp. 1–8, 2009.

[120] N. Dalal, *Finding People in Images and Videos*. PhD thesis, Institut National Polytechnique de Grenoble, July 2006.

[121] M. Ester, H. P. Kriegel, J. Sander, and X. Xu, "A density-based algorithm for discovering clusters in large spatial databases with noise," in *Proc. of the 2nd International Conference on Knowledge Discovery and Data mining*. Portland: AAAI Press, 1996.

[122] G. Karypis, E.-H. Han, and V. Kumar, "Chameleon: Hierarchical clustering using dynamic modeling," *IEEE Computer*, vol. 32, no. 8, pp. 68–75, August 1999.

[123] M. Josiger and K. Kirchner, "Moderne Clusteralgorithmen - Eine vergleichende Analyse auf zweidimensionalen Daten," *Fachgruppentreffen Machinelles Lernen*, pp. 80–84, 2003.

[124] L. Ertoz, M. Steinback, and V. Kumar, "Finding clusters of different sizes, shapes, and density in noisy, high dimensional data," in *Proc. of the 2nd International Conference on Data Mining*, San Francisco (CA), USA, 2003.

[125] J. Sander, M. Ester, H.-P. Kriegel, and X. Xu, "Density-based clustering in spatial databases: The algorithm GDBSCAN and its applications," *Data Mining and Knowledge Discovery*, vol. 2, 1998.

[126] N. Beckmann, H.-P. Kriegel, R. Schneider, and B. Seeger, "The r^*-tree: An efficient and robust access method for points and rectangles," in *ACM SIGMOD International Conference on Management of Data, Atlantic City, NJ*, pp. 322–331, 1990.

[127] P. Steinemann, J. Klappstein, J. Dickmann, H.-J. Wunsche, and F. Hundelshausen, "Determining the outline contour of vehicles in 3D-lidar-measurements," in *Proc. of the IEEE Intelligent Vehicles Symposium*, pp. 479–484, June 2011.

[128] K. Kidono, T. Miyasaka, A. Watanabe, T. Naito, and J. Miura, "Pedestrian recognition using high-definition lidar," in *Proc. of the IEEE Intelligent Vehicles Symposium*, pp. 405–410, June 2011.

[129] DIN 70000, "Straßenfahrzeuge; Fahrzeugdynamik und Fahrverhalten; Begriffe (ISO 8855:1991, modifiziert)," 1994.

[130] L. Krüger, *Model Based Object Classification and Localization in Monocular Images*. PhD Thesis, Universität Bielefeld, Technische Fakultät, 2007.

[131] A. Westenberger, B. Duraisamy, M. Munz, M. Muntzinger, M. Fritzsche, and K. Dietmayer, "Impact of out-of-sequence measurements on the joint integrated probabilistic data association filter for vehicle safety systems," in *Proc. of the IEEE Intelligent Vehicles Symposium*, pp. 438–443, June 2012.

[132] T. Huck, A. Westenberger, M. Fritzsche, T. Schwarz, and K. Dietmayer, "Precise timestamping and temporal synchronization in multi-sensor fusion," in *Proc. of the IEEE Intelligent Vehicles Symposium*, pp. 242–247, June 2011.

[133] R. Lindl, *Tracking von Verkehrsteilnehmern im Kontext von Multisensorsystemen*. PhD Thesis, TU München, 2009.

[134] J.-O. Nilsson and P. Händel, "Time synchronization and temporal ordering of asynchronous measurements of a multi-sensor navigation system," in *Proc. of the Position Location and Navigation Symposium*, 2010.

[135] F. Diewald, J. Klappstein, J. Dickmann, and K. Dietmayer, "Radar-inference-based height estimation of driving environment objects," in *Proc. of the 4th Conference on Safety through Driver Assistance*, April 2010.

[136] I. Haritaoglu, D. Harwood, and L. Davis, "W4: Real-time surveillance of people and their activities," *IEEE Transactions on Pattern Analysis and Machine Intelligence*, vol. 22, no. 8, pp. 809–830, August 2000.

[137] W. Hu, X. Xiao, Z. Fu, D. Xie, T. Tan, and S. Maybank, "A system for learning statistical motion patterns," *IEEE Transactions on Pattern Analysis and Machine Intelligence*, vol. 28, no. 9, pp. 1450–1464, September 2006.

[138] I. Junejo and H. Foroosh, "Trajectory rectification and path modeling for video surveillance," in *Proc. of the IEEE 11th International Conference on Computer Vision*, pp. 1–7, October 2007.

[139] D. Makris and T. Ellis, "Learning semantic scene models from observing activity in visual surveillance," *IEEE Transactions on Systems, Man, and Cybernetics, Part B: Cybernetics*, vol. 35, no. 3, pp. 397–408, June 2005.

[140] J. C. Nascimento, M. A. Figueiredo, and J. S. Marques, "Trajectory modeling using mixtures of vector fields," *Lecture Notes in Computer Science*, vol. 5524, pp. 40–47, 2009.

[141] C. Hermes, C. Wöhler, K. Schenk, and F. Kummert, "Long-term vehicle motion prediction," in *Proc. of the IEEE Intelligent Vehicles Symposium*, pp. 652–657, Xi'an, China, 2009.

[142] E. Käfer, C. Hermes, C. Wöhler, F. Kummert, and H. Ritter, "Recognition and prediction of situations in urban traffic scenarios," in *Proc. of the International Conference on Pattern Recognition*, Istanbul, Turkey, 2010.

[143] C. Keller, C. Hermes, and D. Gavrila, "Will the pedestrian cross? Probabilistic path prediction based on learned motion features," *Pattern Recognition*, vol. 6835, pp. 386–395, 2011.

[144] I. Dagli, G. Breuel, H. Schittenhelm, and A. Schanz, "Cutting-in vehicle recognition for ACC systems- towards feasible situation analysis methodologies," in *Proc. of the IEEE Intelligent Vehicles Symposium*, pp. 925–930, June 2004.

[145] D. Kasper, G. Weidl, T. Dang, G. Breuel, A. Tamke, and W. Rosenstiel, "Object-oriented Bayesian networks for detection of lane change maneuvers," in *Proc. of the IEEE Intelligent Vehicles Symposium*, pp. 673–678, June 2011.

[146] G. S. Aoude, J. Joseph, N. Roy, and J. P. How, "Mobile agent trajectory prediction using bayesian nonparametric reachability trees," in *Proceedings of the American Institute of Aeronautics and Astronautics Infotech*, pp. 1–17, 2011.

[147] C. Laugier, I. Paromtchik, M. Perrollaz, M. Yong, J. Yoder, C. Tay, K. Mekhnacha, and A. Negre, "Probabilistic analysis of dynamic scenes and collision risks assessment to improve driving safety," *IEEE Intelligent Transportation Systems Magazine*, vol. 3, no. 4, pp. 4–19, 2011.

[148] J. Wiest, M. Höffken, U. Kressel, and K. Dietmayer, "Probabilistic trajectory prediction with Gaussian mixture models," in *Proc. of the IEEE Intelligent Vehicles Symposium*, pp. 1162–1167, June 2012.

[149] K. Hayashi, Y. Kojima, K. Abe, and K. Oguri, "Prediction of stopping maneuver considering driver's state," in *Proc. of the IEEE Intelligent Transportation Systems Conference*, pp. 1191–1196, 2006.

[150] T. Huahagen, I. Dengler, A. Tamke, T. Dang, and G. Breuel, "Maneuver recognition using probabilistic finite-state machines and fuzzy logic," in *Proc. of the IEEE Intelligent Vehicles Symposium*, pp. 65–70, June 2010.

[151] R. Tomar, S. Verma, and G. Tomar, "Prediction of lane change trajectories through neural network," in *Proc. of the International Conference on Computational Intelligence and Communication Networks*, pp. 249–253, November 2010.

[152] M. Tsogas, A. Polychronopoulos, N. Floudas, and A. Amditis, "Situation refinement for vehicle maneuver identification and driver intention prediction," in *Proc. of the 10th International Conference on Information Fusion*, pp. 1–8, December 2007.

[153] A. Polychronopoulos, M. Tsogas, A. Amditis, and L. Andreone, "Sensor fusion for predicting vehicles' path for collision avoidance systems," *IEEE Transactions on Intelligent Transportation Systems*, vol. 8, no. 3, pp. 549–562, September 2007.

[154] C. Zong, C. Wang, D. Yang, and H. Yang, "Driving intention identification and maneuvering behavior prediction of drivers on cornering," in *Proc. of the International Conference on Mechatronics and Automation*, pp. 4055–4060, December 2009.

[155] T. Taniguchi, S. Nagasaka, K. Hitomi, N. P. Chandrasiri, and T. Bando, "Semiotic prediction of driving behavior using unsupervised double articulation analyzer," in *Proc. of the IEEE Intelligent Vehicles Symposium*, pp. 849–854, June 2012.

[156] M. Liebner, M. Baumann, F. Klanner, and C. Stiller, "Driver intent inference at urban intersections using the intelligent driver model," in *Proc. of the IEEE Intelligent Vehicles Symposium*, pp. 1162–1167, June 2012.

[157] P. Evans, "Finding common subsequences with arcs and pseudoknots," in *Combinatorial Pattern Matching*, M. Crochemore and M. Paterson, Eds., ser. Lecture Notes in Computer Science, vol. 1645, pp. 270–280. Springer, Berlin Heidelberg, 1999.

[158] S. Bereg, M. Kubica, T. Walen, and B. Zhu, "RNA multiple structural alignment with longest common subsequences," *Journal of Combinatorial Optimization*, vol. 13, no. 2, pp. 179–188, February 2007.

[159] H. Fashandi and E. A. M. Moghaddam, "A new rotation invariant similarity measure for trajectories," in *Proc. of the IEEE International Symposium on Computational Intelligence in Robotics and Automation*, pp. 631–634, June 2005.

[160] W. Bailer, "Eliminating the back-tracking step in the longest common subsequence (LCSS) algorithm for video sequence matching," in *Proc. of the International Conference on Semantic and Digital Media Technology*, Koblenz, December 2008.

[161] J. W. Hunt and T. G. Szymanski, "A fast algorithm for computing longest common subsequences," *Programming Techniques*, vol. 20, no. 5, pp. 350–353, May 1977.

[162] C. S. Iliopoulos, M. Kubica, M. S. Rahman, and T. Walen, "Algorithms for computing the longest parameterized common subsequence," in *Lecture Notes in Computer Science*, B. Ma and K. Zhang, Eds., ser. Combinatorial Pattern Matching, vol. 1580, pp. 265–273. Springer, Berlin Heidelberg, 2007.

[163] T. Jiang, G.-H. Lin, B. Ma, and K. Zhang, "The longest common subsequence problem for arc-annotated sequences," in *Lecture Notes in Computer Science*, R. Giancarlo and S. Sankoff, Eds., ser. Combinatorial Pattern Matching, vol. 1848, pp. 154–165. Springer, Berlin Heidelberg, 2000.

[164] J. Kloetzli, B. Strege, J. Decker, and M. Olano, "Parallel longest common subsequence using graphics hardware," in *Proc. of the Eurographics Symposium on Parallel Graphics and Visualization*, 2008.

[165] J. Uebersax, "Genetic counseling and cancer risk modeling: An application of Bayes nets," *Methodology Division Research Report*, August 2004.

[166] N. Friedman, M. Linial, I. Nachman, and D. Pe'er, "Using Bayesian networks to analyze expression data," *Journal of Computational Biology*, vol. 7, no. 3-4, pp. 601–620, July 2004.

[167] E. Dagan, O. Mano, G. Stein, and A. Shashua, "Forward collision warning with a single camera," in *Proc. of the IEEE Intelligent Vehicles Symposium*, pp. 37–42, June 2004.

[168] J. Hillenbrand, A. Spieker, and K. Kroschel, "Efficient decision making for a multi-level collision mitigation system," in *Proc. of the IEEE Intelligent Vehicles Symposium*, pp. 460–465, 2006.

[169] R. Labayrade, C. Royere, and D. Aubert, "A collision mitigation system using laser scanner and stereovision fusion and its assessment," in *Proc. of the IEEE Intelligent Vehicles Symposium*, pp. 441–446, June 2005.

[170] A. Vahidi and A. Eskandarian, "Research advances in intelligent collision avoidance and adaptive cruise control," *IEEE Transactions on Intelligent Transportation Systems*, vol. 4, pp. 143–153, 2003.

[171] J. Hillenbrand, A. Spieker, and K. Kroschel, "A multilevel collision mitigation approach: Its situation assessment, decision making, and performance tradeoffs," *IEEE Transactions on Intelligent Transportation Systems*, vol. 7, no. 4, pp. 528–540, December 2006.

[172] A. Polychronopoulos, M. Tsogas, A. Amditis, U. Scheunert, L. Andreone, and F. Tango, "Dynamic situation and threat assessment for collision warning systems: The EUCLIDE approach," in *Proc. of the Intelligent Vehicles Symposium*, pp. 636–641, June 2004.

[173] A. Tamke, T. Dang, and G. Breuel, "A flexible method for criticality assessment in driver assistance systems," in *Proc. of the IEEE Intelligent Vehicles Symposium*, pp. 697–702, June 2011.

[174] E. Coelingh, A. Eidehall, and M. Bengtsson, "Collision warning with full auto brake and pedestrian detection – A practical example of automatic emergency braking," in *Proc. of the 13th International IEEE Conference on Intelligent Transportation Systems*, pp. 155–160, September 2010.

[175] L. Yang, J. H. Yang, E. Feron, and V. Kulkarni, "Development of a performance-based approach for a rear-end collision warning and avoidance system for automobiles," in *Proc. of the IEEE Intelligent Vehicles Symposium*, pp. 316–321, June 2003.

[176] J. Jansson, *Collision Avoidance Theory with Applications to Automotive Collision Mitigation*. PhD Thesis 950, Department of Electrical Engineering, University of Linköping, 2005.

[177] A. Berthelot, A. Tamke, T. Dang, and G. Breuel, "Handling uncertainties in criticality assessment," in *Proc. of the IEEE Intelligent Vehicles Symposium*, pp. 571 –576, June 2011.

[178] ——, "A novel approach for the probabilistic computation of time-to-collision," in *Proc. of the IEEE Intelligent Vehicles Symposium*, pp. 1173–1178, June 2012.

[179] C. Rodemerk, S. Habenicht, A. Weitzel, H. Winner, and T. Schmitt, "Development of a general criticality criterion for risk estimation of driving situations and its application to a maneuver-based lane change assistance system," in *Proc. of the IEEE Intelligent Vehicles Symposium*, pp. 264–269, June 2012.

[180] S. Lefèvre, C. Laugier, and J. Ibanez-Guzmán, "Risk assessment at road intersections: Comparing intention and expectation," in *Proc. of the IEEE Intelligent Vehicles Symposium*, pp. 165–171, June 2012.

[181] A. Ferrara and C. Vecchio, "Second order sliding mode control of vehicles with distributed collision avoidance capabilities," *Mechatronics*, vol. 19, no. 4, pp. 471–477, 2009.

[182] C. Wakim, S. Capperon, and J. Oksman, "A Markovian model of pedestrian behavior," in *Proc. of the IEEE International Conference on Systems, Man, and Cybernetics*, pp. 4028–4033, October 2004.

[183] F. Large, D. Vasquez, T. Fraichard, and C. Laugier, "Avoiding cars and pedestrians using velocity obstacles and motion prediction," in *Proc. of the IEEE Intelligent Vehicles Symposium*, pp. 375–379, June 2004.

[184] K. Smith and J.-E. Källhammer, "Field of safe travel: Using location and motion information to increase driver acceptance of pedestrian alerts," in *Proc. of the IEEE Intelligent Vehicles Symposium*, pp. 1168–1172, June 2012.

[185] A. Broadhurst, S. Baker, and T. Kanade, "Monte Carlo road safety reasoning," in *Proc. of the IEEE Intelligent Vehicles Symposium*, pp. 319–324, June 2005.

[186] A. Eidehall and L. Petersson, "Threat assessment for general road scenes using monte carlo sampling," in *Proc. of the IEEE Intelligent Transportation Systems Conference*, pp. 1173–1178, September 2006.

[187] S. Danielsson, L. Petersson, and A. Eidehall, "Monte Carlo based threat assessment: Analysis and improvements," in *IEEE Intelligent Vehicles Symposium*, pp. 233 –238, June 2007.

[188] J. Jansson and F. Gustafsson, "A framework and automotive application of collision avoidance decision making," *Automatica*, vol. 44, no. 9, pp. 2347–2351, 2008.

[189] J. Hu, M. Prandini, and S. Sastry, "Probabilistic safety analysis in three dimensional aircraft flight," in *Proc. of the 42nd IEEE Conference on Decision and Control*, pp. 5335–5340, December 2003.

[190] ——, "Aircraft conflict prediction in the presence of a spatially correlated wind field," *IEEE Transactions on Intelligent Transportation Systems*, vol. 6, no. 3, pp. 326–340, September 2005.

[191] A. E. Broadhurst, S. Baker, and T. Kanade, "A prediction and planning framework for road safety analysis, obstacle avoidance and driver information," in *Proc. of the 11th World Congress on Intelligent Transportation Systems*, Pittsburgh (PA), USA, October 2004.

[192] A. Eidehall and L. Petersson, "Statistical threat assessment for general road scenes using monte carlo sampling," *IEEE Transactions on Intelligent Transportation Systems*, vol. 9, pp. 137–147, March 2008.

[193] H. Zhang, "The optimality of naive bayes," in *Proc. of the Florida Artificial Intelligence Research Society Conference*, 2004.

[194] M. Rosenblatt, "Remarks on some nonparametric estimates of a density function," *The Annals of Mathematical Statistics*, vol. 27, pp. 832–837, 1956.

[195] A. W. Bowman and A. Azzalini, *Applied Smoothing Techniques for Data Analysis*. Oxford University Press, New York, 1997.

[196] P. P. Dey, S. Chandra, and S. Gangopadhaya, "Lateral distribution of mixed traffic on two-lane roads," *Journal of Transportation Engineering*, vol. 132, pp. 597–600, 2006.

[197] E. Velenis and P. Tsiotras, "Optimal velocity profile generation for given acceleration limits: theoretical analysis," in *Proceedings of the American Control Conference*, pp. 1478–1483, 2005.

[198] Y. Dodge, *The Oxford Dictionary of Statistical Terms*. Oxford University Press, 2003.

[199] A. Ghaleb, L. Vignaud, and J. Nicolas, "Micro-Doppler analysis of wheels and pedestrians in ISAR imaging," *Signal Processing, IET*, vol. 2, pp. 301–311, 2008.

[200] F. Klanner, *Entwicklung eins kommunikationsbasierten Querverkehrsassistenten im Fahrzeug.* PhD Thesis, Technische Universität Darmstadt, 2008.

[201] J. Hillenbrand, *Fahrerassistenz zur Kollisionsvermeidung.* PhD Thesis, Universität Karlsruhe, 2007.

[202] K.-C. Chang and Y. Bar-Shalom, "Joint probabilistic data association for multitarget tracking with possibly unresolved measurements," in *Proc. of the American Control Conference*, pp. 466–471, 1983.

[203] S. E. Lyshevski, *MEMS and NEMS: Systems, Devices and Structures.* CRC Press LLC, USA, 2002.

[204] H. Winner, *Handbuch Fahrerassistenzsysteme.* Vieweg+Teubner – Springer Fachmedien Wiesbaden GmbH, 2012.

[205] O. Kosheleva, V. Kreinovich, G. Mayer, and H. T. Nguyen, "Computing the cube of an interval matrix is NP-hard," in *Proc. of the ACM Symposium on Applied Computing*, pp. 1449–1453, 2005.

[206] L. Jaulin, M. Kieffer, and O. Didrit, *Applied Interval Analysis.* Springer, 2006.

List of Publications

[207] C. Otto, W. Gerber, F. Puente León, and J. Wirnitzer, "A joint integrated probabilistic data association filter for pedestrian tracking across blind regions using monocular camera and radar," in *Proc. of the IEEE Intelligent Vehicles Symposium*, June 2012.

[208] C. Otto, F. Puente León, and J. Wirnitzer, "Long-term trajectory classification and prediction of commercial vehicles for the application in advanced driver assistance systems," in *Proc. of the American Control Conference*, June 2012.

[209] C. Otto, F. Puente León, and A. Schwarzhaupt, "A strategy on situation evaluation for driver assistance systems in commercial vehicles considering pedestrians in urban traffic," in *Proc. of the IEEE Intelligent Vehicles Symposium*, June 2011.

[210] C. Otto, A. Schwarzhaupt, and F. Puente León, "Erreichbarkeitsmengenanalyse für die Situationsbewertung in Fahrerassistenzsystemen der aktiven Sicherheit im Nutzfahrzeug," in *Proc. of the VDI Nutzfahrzeugetagung*, May 2011.

[211] C. Otto, D. Keller, G. Seemann, and O. Dössel, "Integrating beta-adrenergic signaling into a computational model of human cardiac electrophysiology," in *Proc. of the World Congress on Medical Physics and Biomedical Engineering, IFMBE*, vol. 25, pp. 1033–1036, October 2009.

[212] C. Otto, U. Wiesel, J. Wirnitzer, A. Schwarzhaupt, and F. Puente León, "Cost-efficient environment perception using multiple sensors for driver assistance systems in commercial vehicles," *In Fernando Puente and Klaus Dostert editors, Reports on Industrial Information Technology, KIT Scientific Publishing, Karlsruhe*, vol. 12, pp. 181–191, 2010.

List of Patents

[213] C. Otto, A. Schwarzhaupt, I. Vegh, U. Wiesel, and J. Wirnitzer, "Verfahren zum Betrieb eines Fahrzeugs," German Patent DE 102 010 033 774A1, May 12, 2011.

[214] ——, "Bilderfassungsvorrichtung für ein Fahrzeug und Verfahren zum Betrieb einer Bilderfassungsvorrichtung," German Patent DE 102 010 023 199A1, February 10, 2011.

[215] ——, "Konzept für eine Lagewinkelkorrektur für kamerabasierte Fahrerassistenzsysteme," German Patent DE 1 102 010 048 143, July 28, 2011.

[216] C. Otto, A. Schwarzhaupt, and W. Stahl, "Verfahren zum Erstellen und zur Nutzung einer Gefahrenkarte eines Kraftfahrzeug-Navigationssystems," German Patent DE 102 010 055 370A1, June 21, 2012.

[217] C. Otto, A. Schwarzhaupt, and J. Wirnitzer, "Verfahren und System zur Vorhersage von Kollisionen," German Patent DE 102 011 010 864A8, December 08, 2011.

[218] ——, "Verfahren zur Assistenz eines Fahrers bei Spurwechseln und Spurwechselassistenzsystem," German Patent DE 102 010 054 221A1, August 25, 2011.

[219] C. Otto, A. Schwarzhaupt, I. Vegh, U. Wiesel, and J. Wirnitzer, "Verfahren zur Objekterkennung mittels Bilddaten," German Patent DE 102 011 013 768A1, November 10, 2011.

[220] ——, "Verfahren zur Erfassung einer Umgebung eines Fahrzeugs," German Patent DE 102 010 032 063A1, May 12, 2011.

List of Supervised Theses

[221] D. Penning, *Vermessung einer Fußgängerdetektion durch Radar und Monokularkamera zur Parametrierung eines Joint-Integrated-Probabilistic-Data-*

Association-Filters. Bachelor thesis, Karlsruhe Institute of Technology, Institute of Industrial Information Technology, November 2011.

[222] W. Gerber, *Vergleich von EKF und JIPDA-Filter zur Fußgängerverfolgung im Fahrzeug-Umfeld*. Diploma thesis, Karlsruhe Institute of Technology, Institute of Industrial Information Technology, March 2012.

[223] M. Biachi Dambros, *Ansatz zur Manöverprädiktion und Pfadprädiktion eines LKWs für die Anwendung in der Situationsbewertung von Fahrerassistenzsystemen*. Student research project, Karlsruhe Institute of Technology, Institute of Industrial Information Technology, January 2013.

[224] I. Thomas, *Analyse des Bewegungsverhaltens von Fußgängern im Straßenverkehr*. Bachelor thesis, Duale Hochschule Baden-Württemberg, Campus Friedrichshafen, September 2012.

[225] J. Bräuler, *Entwicklung einer sensordatenfusionierenden Simulationsumgebung für den Einsatz in umfelderfassenden Assistenzsystemen im Nutzfahrzeug*. Diploma thesis, Hochschule Darmstadt, March 2010.

[226] M. Kolditz, *Entwicklung einer Spur-Objekt-Fusion auf Basis von Radar- und Mono-Kamera-Daten für Fahrerassistenzsysteme im Nutzfahrzeug*. Diploma thesis, Karlsruhe Institute of Technology, Institute of Product Engineering, April 2010.

[227] T. Heger, *Bewertung und Vergleich unterschiedlicher Sattelzug-Fahrdynamikmodelle zur Unterstützung der Situationsbewertung für Fahrerassistenzsysteme*. Bachelor thesis, Duale Hochschule Baden-Württemberg, Campus Mannheim, 2010.

[228] M. Kleebaur, *Fahrzeugintegration eines Objekt-zu-Spur-Fusionsansatzes auf Basis von Radar- und Mono-Kamera-Daten als Grundlage für Fahrerassistenzsysteme*. Bachelor thesis, Hochschule Aalen, 2010.

[229] M. Couderc, *Optimierung und Validierung einer Objekt-zu-Spur-Fusion auf Basis von Radar- und Mono-Kamera-Daten für Fahrerassistenzsysteme im Nutzfahrzeug*. Master thesis, Hochschule Ulm, 2011.

[230] T. Nguyen, *Erreichbarkeitsmengenanalyse zur Unterstützung der Überwachung des toten Winkels eines Nutzfahrzeugs*. Diploma thesis, Karlsruhe Institute of Technology, Institute of Industrial Information Technology, February 2012.

[231] A. Pflug, *Entwicklung eines Werkzeugs zur synchronen Darstellung der Daten von Kameramodul, Radar und Laser-Scanner für die Entwicklung umfelderfassender Fahrerassistenzsysteme*. Bachelor thesis, Fachhochschule Südwestfalen, January 2012.

[232] S. Schelle, *Klassifikation und Prädiktion von Trajektorien*. Diploma thesis, Universität Stuttgart, 2012.

**Forschungsberichte
aus der Industriellen Informationstechnik
(ISSN 2190-6629)**

Institut für Industrielle Informationstechnik
Karlsruher Institut für Technologie (KIT)

Hrsg.: Prof. Dr.-Ing. Fernando Puente León, Prof. Dr.-Ing. habil. Klaus Dostert

Die Bände sind unter www.ksp.kit.edu als PDF frei verfügbar oder als Druckausgabe bestellbar.

Band 1   Pérez Grassi, Ana
         **Variable illumination and invariant features for detecting and
         classifying varnish defects.** (2010)
         ISBN 978-3-86644-537-6

Band 2   Christ, Konrad
         **Kalibrierung von Magnet-Injektoren für Benzin-Direkteinspritzsysteme
         mittels Körperschall.** (2011)
         ISBN 978-3-86644-718-9

Band 3   Sandmair, Andreas
         **Konzepte zur Trennung von Sprachsignalen in unterbestimmten
         Szenarien.** (2011)
         ISBN 978-3-86644-744-8

Band 4   Bauer, Michael
         **Vergleich von Mehrträger-Übertragungsverfahren und Entwurfs-
         kriterien für neuartige Powerline-Kommunikationssysteme zur
         Realisierung von Smart Grids.** (2012)
         ISBN 978-3-86644-779-0

Band 5   Kruse, Marco
         **Mehrobjekt-Zustandsschätzung mit verteilten Sensorträgern am
         Beispiel der Umfeldwahrnehmung im Straßenverkehr** (2013)
         ISBN 978-3-86644-982-4

Band 6   Dudeck, Sven
         **Kamerabasierte In-situ-Überwachung gepulster
         Laserschweißprozesse** (2013)
         ISBN 978-3-7315-0019-3

Band 7   Liu, Wenqing
         **Emulation of Narrowband Powerline Data Transmission Channels
         and Evaluation of PLC Systems** (2013)
         ISBN 978-3-7315-0071-1

Forschungsberichte aus der Industriellen Informationstechnik (ISSN 2190-6629)

Institut für Industrielle Informationstechnik | Karlsruher Institut für Technologie (KIT)

Hrsg.: Prof. Dr.-Ing. Fernando Puente León, Prof. Dr.-Ing. habil. Klaus Dostert

Band 8   Otto, Carola
              **Fusion of Data from Heterogeneous Sensors with Distributed
              Fields of View and Situation Evaluation for Advanced Driver
              Assistance Systems.** (2013)
              ISBN 978-3-7315-0073-5